The Political Economy of Climate Change Adaptation

The Political Economy of Climate Change Adaptation

Benjamin K. Sovacool
Professor of Business and Social Sciences, Aarhus University, Denmark
Professor of Energy Policy, Science Policy Research Unit, University of Sussex, United Kingdom

and

Björn-Ola Linnér
Professor of Environmental Change, Linköping University, Sweden
Associate Fellow, Institute for Science, Innovation and Society, Oxford University, United Kingdom

palgrave macmillan

nature climate change

First published 2016 by
PALGRAVE MACMILLAN

Palgrave Macmillan in the UK is an imprint of Macmillan Publishers Limited, registered in England, company number 785998, of Houndmills, Basingstoke, Hampshire RG21 6XS.

Palgrave Macmillan in the US is a division of St Martin's Press LLC, 175 Fifth Avenue, New York, NY 10010.

Palgrave Macmillan is the global academic imprint of the above companies and has companies and representatives throughout the world.

Palgrave® and Macmillan® are registered trademarks in the United States, the United Kingdom, Europe and other countries.

ISBN: 978–1–137–49672–0

This book is printed on paper suitable for recycling and made from fully managed and sustained forest sources. Logging, pulping and manufacturing processes are expected to conform to the environmental regulations of the country of origin.

A catalogue record for this book is available from the British Library.

Library of Congress Cataloging-in-Publication Data

Sovacool, Benjamin K., author.
The political economy of climate change adaptation / Benjamin K. Sovacool, Björn-Ola Linnér.
pages cm
ISBN 978–1–137–49672–0 (hardback)
1. Climatic changes – Political aspects. 2. United Nations Framework Convention on Climate Change (1992 May 9) 3. Climatic changes – Government policy. I. Linnér, Björn-Ola, author. II. Title.

QC903.S627 2015
363.738′74—dc23 2015021813

Contents

List of Figures

List of Tables

Preface and Acknowledgments

Writing in 1886 about the scientific method, G.K. Gilbert once argued that "In the testing of hypotheses lies the prime difference between the investigator and the theorist. The one seeks diligently for the facts which may overthrow his or her tentative theory, the other closes his or her eyes to these and searches only for those which will sustain it."[1] This book was particularly difficult to research and to write because our project blended the roles of investigator and theorist together. That is, we did start with a theory of the political economy of adaptation in mind (what we've called "the 4Es" of enclosure, exclusion, encroachment, and entrenchment) and sought as many examples that we could find to support it. However, unlike a truly zealous theorist, we did not discount or ignore contravening evidence; we only tried to deepen our analysis to explain and contextualize it. We thus believe that our theoretical framework, although it was deductive rather than inductive, was empirically driven.

Nor, of course, did this project spring from a vacuum. We have both been working on the topic of climate adaptation for years, though not always using a political economy framework. We sensed the lack of a political economy discourse in adaptation projects as we came to learn more about how adaptation was being implemented around the world. Most notably, in The Political Economy of Climate Adaptation, a commentary published in *Nature Climate Change* in July 2015 (doi: 10.1038/nclimate26655), we identified specific political economy elements – enclosure, exclusion, encroachment and entrenchment (4Es) – in a series of case studies including the Wonthaggi desalination plant in Australia, disaster recovery in Honduras, coastal protection in Norway, sea barriers in Alaska, marine protected areas in Tanzania, climate-proofing infrastructure in the Maldives, livelihood diversification in Burkina Faso, and disaster relief in Kenya. Feeling that the topic deserved more in-depth treatment, we expanded that work in this book to look at instances where the 4Es occur not as singular, isolated events but in synergy and simultaneously, in a sort of feedback loop. To this end, we explore here four different case studies: the displacement of char communities in Bangladesh, the Dutch Delta Works in the Netherlands, Hurricane Katrina reconstruction efforts in the United States, and the politics of technology transfer and knowledge inequality within the United Nations Framework Convention on Climate Change.

The credit for this book, moreover, does not lie with the authors alone. The authors are extremely grateful to the Rockefeller Foundation for Grant RES2009315, to the Norden Top-level Research Initiative sub-programme "Effect Studies and Adaptation to Climate Change" for the Nordic Centre of Excellence for Strategic Adaptation Research (NORD-STAR) project, and to the Research Council Formas for Grant 2013-1557 "Identifying Thresholds for Maladaptation in Nordic Agriculture", which have supported elements of the work reported here.

Just as important, we thank Dr. Sarah Ryan from Yale Law School for considerately reviewing the entire book, and participants at the "Political Economy of Energy" Catalytic Research Workshop of the 56th Annual Convention of the International Studies Association, New Orleans, Louisiana (2015) for helpful suggestions. We are also indebted to numerous colleagues and peers who took the time to review and comment on particular chapter drafts.

In that vein, Kate J. Neville and Erika Weinthal from Duke University, Michael Evan Goodsite from the University of Southern Denmark and Sofie Storbjörk, Linköping University, provided thoughtful suggestions for how to tighten our conceptual framework in Chapter 1.

Mohammed Mamun Or Rashid from the USAID's Climate Resilient Ecosystems and Livelihoods (CREL) Project and Shapan Adnan from the University of Oxford offered sage advice concerning Chapter 2's case study on Bangladesh.

Richard J.T. Klein from the Stockholm Environment Institute, Wiebe E. Bijker from Maastricht University, Job Dronkers from Deltares, and Rob van Ginkel from the University of Amsterdam offered outstanding suggestions for the revision of Chapter 4's case study on the Dutch Delta Works.

Charles Richard Baker from Adelphi University, Lynn Weber from the University of South Carolina, and Lori Peek from Colorado State University graciously took the time to review and comment on Chapter 5's case study of Hurricane Katrina.

David Victor from the University of California at San Diego, Anna Bohman from Linköping University, Åsa Persson from Stockholm Environment Institute and Thomas Hale from Oxford University provided helpful suggestions for revisions to Chapter 5's case study on the UNFCCC.

Sirkku Juhola of Helsinki University and Julie Wilk, Mattias Hjerpe, and Erik Glaas from Linköping University offered insights into Chapter 6's recommendations.

Of course, despite this assistance, any opinions, findings, and conclusions or recommendations expressed in this material are those of the authors alone and do not necessarily reflect the views of the Rockefeller Foundation, NORD-STAR, Formas or our peer reviewers.

Lastly, both the authors wish to thank their families for the encouragement and support needed to finish a book like this.

Note

1. Gilbert, G.K. (1886). On the validity of scientific hypotheses. *American Journal of Science*, 31, 22.

1
Introduction to the Political Economy of Climate Change Adaptation

1 Introduction

In 2007, Percy Schmeiser, an elderly farmer from Saskatchewan, Canada, unknowingly harvested a crop of canola that contained a herbicide-resistant gene patented by one of the world's largest agricultural biotechnology companies, Monsanto.[1] Schmeiser claimed that the canola had sprouted from seeds that had blown off passing trucks or spread from adjacent fields and mistakenly taken root on his farm – meaning he was not liable for patent infringement. Monsanto disagreed, and the case went all the way to the Canadian Supreme court, which reached a "bizarre" 5–4 decision that Schmeiser had infringed upon Monsanto's patents, but was not liable for any damages.[2]

The implications of this anecdote are manifold and prophetic, extending well beyond an instance of a small-scale farmer, a "David," losing to a large transnational corporation, a "Goliath." It is an instance where Monsanto was in the process of actively modifying the genetic makeup of plants so that they were better adapted to the environment. It is an example of the commodification and corporatization of nature, where a company has patented and engineered a type of biodiversity. It is, lastly, a case where the profit motives of Monsanto pushed them to sue a farmer who likely only used their seeds by accident. In short, it is a case that raises a difficult question: is it appropriate to adapt or change a piece of nature, and to exclude others from its use, in order to make money or serve a particular interest and if so under what circumstances?

This book argues that in some cases of climate change adaptation – altering infrastructure, institutions, or ecosystems to respond to the impacts of climate change – profiteering, exclusion, or serving particular entrenched interests is neither appropriate nor effective. Notwithstanding the great promise that climate change adaptation efforts offer society, something might be going softly, silently awry with them. A survey of hundreds of studies of climate change adaptations implemented over the past decade reached a worrying conclusion: many projects were not helping the most vulnerable, and were instead strengthening established sectors that had already received large shares of adaptation funding.[3] Another article warned that within adaptation projects, "deeply entrenched social institutions and norms may influence which group members will be able to have a voice and ultimately exercise rights."[4] Similarly, Biermann, Pattberg, and Zelli demonstrated how adaptation interventions are geared to serve particular interests, be it donor agencies or the agendas of particular companies.[5] Adger hypothesized that "vulnerability to environmental change does not exist in isolation from the wider political economy of resource use" and that "vulnerability is driven by inadvertent or deliberate human action that reinforces self-interest and the distribution of power in addition to interacting with physical and ecological systems."[6]

Notwithstanding these statements, little research has explored the systematic or structural aspects of the politics of adaptation head on, or the historical failures associated with adaptation projects. Instead, researchers have attempted to untangle separate threads of the topic sporadically. We group this prior research into seven strands of focus on the political economy aspects of adaptation:

1. Corruption in climate change adaptation and the politics of lobbying;[7,8,9]
2. Maladaptation, where adaptation projects unintentionally lower resilience or increase Greenhouse Gas (GHG) emissions;[10,11]
3. The winners and losers of climate change (e.g., who gets longer growing seasons compared to who suffers drought);[12]
4. The "double inequity" between responsibility for climate change (large industrialized emitters) and vulnerability to it (small developing economies);[13]
5. Sustainable adaptation, homing in on the consequences of adaptation policies and measures for other sustainable development goals;[14]
6. Tradeoffs between mitigation (reducing emissions) and adaptation (coping with consequences);[15,16,17,18]

7. Climate change and adaptation as a discourse[19,20] – what Taylor calls an "array of discursive coordinates and institutional practices"[21] – that serves to homogenize perspectives and diminish the autonomy of outsiders.

No research, as of yet, has systematically explored the tradeoffs within or between adaptation projects, or provided explanations for *how* the political economy of adaptation might operate. And so, two research questions emerge: Who are the winners and losers of adaptation projects? What underlying processes might affect the inequitable distribution of adaptation costs and benefits?

To provide an answer, this book documents the presence of four attributes to adaptation projects – processes we have termed enclosure, exclusion, encroachment, and entrenchment – cutting across the economic, political, ecological, and social dimensions summarized by Table 1.1. In a contribution to *Nature Climate Change* we pointed at eight other examples of these four processes.[22] In this book, we find the four processes at work simultaneously in at least four case studies, two of them focusing on developing countries, two on developed countries. These cases involve the displacement of *char* communities in Bangladesh, the Dutch Delta Works in the Netherlands, Hurricane Katrina reconstruction efforts in the United States, and the politics of technology transfer

Table 1.1 Conceptual typology of enclosure, exclusion, encroachment, and entrenchment

Concept	Dimension	Explanation
Enclosure	Economic	Capturing resources or authority: transferring public assets into private hands, or the expansion of private roles into the public sector
Exclusion	Political	Marginalizing stakeholders: limiting access to adaptation decision-making processes and fora
Encroachment	Ecological	Damaging the environment: intruding on biodiversity areas or other areas with predisposed land uses, or interfering with ecosystem services
Entrenchment	Social	Worsening inequality: aggravating the disempowerment of women or minorities and/or worsening concentrations of wealth

Source: Authors.

and knowledge inequality within the United Nations Framework Convention on Climate Change (UNFCCC). Our book concludes with a discussion of the broader implications of the political economy of adaptation for analysts, program managers, and climate researchers at large. In sum, the politics of adaptation must be taken into account so that projects can maximize their efficacy and avoid marginalizing those most vulnerable to the impacts of climate change.

This introductory chapter begins by briefly familiarizing readers with the basics of climate change adaptation, defining its core elements, and offering a series of compelling reasons why it matters to society. It then summarizes four separate fields of inquiry – political economy, political ecology, justice theory, and critical development studies – that suggest we view adaptation not so much as a technical and economic process of coping with climate change, but as a deeply political and ethical process. The chapter then introduces the key theoretical framework of the book, namely the processes of enclosure, exclusion, encroachment, and entrenchment, and discusses the book's primary research methods, before previewing the chapters to come.

Now, one quick, central caveat before we embark upon our textual odyssey. Before we give readers the wrong impression, neither of us is actually against human investments in climate change adaptation. We support adaptation projects and processes in both principle and practice. Many, if not most, projects probably have positive results. The four case studies examined here are a bit more complex, and may not produce net social benefits, but our main thesis is not that adaptation ought to be rejected. On the contrary. Instead, we argue that adaptation analysts and advocates need to become more cognizant of some of the potential downsides to adaptation, and take political economy into account to ensure that projects are freely decided upon, benefits fairly distributed and thus ultimately more effective in creating resilience for those vulnerable to a changing climate.

2 The basics of adaptation

It is best to start with a few definitions.

2.1 Adaptation

The concept of adaptation was defined in the Intergovernmental Panel on Climate Change's (IPCC) Fourth Assessment Report as "adjustment in natural or human systems in response to actual or expected climatic stimuli or their effects, which moderates harm or exploits beneficial opportunities."[23] Adaptation is often contrasted with mitigation, which the report defines as "an anthropogenic intervention to reduce the anthropogenic forcing of the climate system; it includes strategies to reduce greenhouse gas sources and emissions and enhance greenhouse gas sinks."[24] Put in very simple terms, mitigation is avoiding climate change, whereas adaptation is coping with climate change, or as Brown and Sovacool put it, mitigation is "avoiding the unmanageable" whereas adaptation is "managing the unavoidable."[25] That is the simple version. In practice, adaptation can be understood in a variety of ways. Table 1.2 introduces a number of related, though differing, prominent definitions of adaptation.

Though these definitions tell us that adaptation involves how humans deal with climate change, they do not tell us *how* adaptation usually occurs. One way of categorizing adaptation projects or interventions is by sector. As Table 1.3 indicates, climate-proofing buildings is one critical infrastructural option, along with hardening seawalls and other seacoast structures against rises in sea level. Institutional options tend to focus on improving the governance or public management of climate-related risks, such as the creation of early warning systems for severe weather events. Community options focus on the vitality and well-being of people, and can include risk transfer or insurance measures addressing low-frequency, high-severity weather events such as once-in-100-year floods.[26] Lastly, ecosystems and natural capital – forests, wetlands, crops – can be made more adaptive.

Adaptation can be either reactive, seeking to blunt impacts that have happened, or proactive in the sense that it strategically anticipates and strives to avoid the effects of future undesired changes. Adaptation interventions can also be classified by their underlying logic or type. Ahmed, Alam, and Rahman categorized adaptation measures into at least seven strategies.[27] The first is to "bear losses," which amounts to basically doing nothing. John Holden in the US called this the "suffering" response to climate change.[28] A second is to "share losses," to provide subsidies to people living in vulnerable areas, or to increase taxes for those living in protected areas. Other efforts falling under this option include offering crop loss insurance or homestead insurance, or providing relief or credit for rehabilitation. A third is to "modify the threat," switching to alternative cropping patterns, improving early warning systems, or erecting breakwaters in an island to safeguard tourism. A fourth is to "prevent the effect," such as building embankments to prevent flooding, or constructing flood control infrastructure. A fifth is to "change use," such as switching from rice to growing shrimp or aquaculture in areas suffering saltwater intrusion. A sixth is to

Table 1.2 Eight definitions of climate change adaptation

Definition	Year	Source
"The process through which people reduce the adverse effects of climate on their health and wellbeing, and take advantage of the opportunities that their climatic environment provides"	1992	Canadian Climate Center[a]
"Adjustments to enhance the viability of social and economic activities and to reduce their vulnerability to climate, including its current variability and extreme events as well as longer-term climate change"	1993	University of Guelph[b]
"Any adjustment, whether passive, reactive or anticipatory, that is proposed as a means for ameliorating the anticipated adverse consequences associated with climate change"	1993	US Army Corps of Engineers[c]
"The degree to which adjustments are possible in practices, policies, or structures of systems to projected or actual changes of climate"	1996	Intergovernmental Panel on Climate Change Second Assessment[d]
"All adjustments in behavioral or economic structure that reduce the vulnerability of society to changes in the climate system"	1996	Smith et al. 1996[e]
"Improving country resilience against climate risks"	2011	Organization of Economic Co-Operation and Development[f]
"A range of approaches to address loss and damage associated with the adverse effects of climate change, including impacts related to extreme weather events and slow onset events"	2014	United Nations Framework Convention on Climate Change[g]

Notes: [a] Burton, I. (1992). *Adapt and Thrive*. Ontario, Canada: Canadian Climate Center.

[b] Smit, B. (1993). Adaptation to Climate Variability and Change. Environmental Paper No. 19. Ontario, Canada: University of Guelph.

[c] Stakhiv, E. (1993). *Evaluation of IPCC Adaptation Strategies: Draft Report*. Fort Belvoir, VA: US Army Corps of Engineers.

[d] Watson, R.T., Zinyowera, M.C., & Moss, R.H. (eds) (1996). *Climate Change 1995: Impacts, Adaptations and Mitigation of Climate Change: Scientific-Technical Analyses*. Cambridge, UK: Cambridge University Press.

[e] Smith, J.B., et al. (eds) (1996). *Adapting to Climate Change: An International Perspective*. New York, NY: Springer-Verlag.

[f] Organisation for Economic Co-operation and Development (2011). *Adaptation to climate change*. Available at: http://www.oecd.org/env/cc/adaptation.htm.

[g] United Nations Framework Convention on Climate Change (2014). *Adaptation: Thematic areas of the work programme on loss and damage*. Available at: http://unfccc.int/adaptation/workstreams/loss_and_damage/items/7546.php.

Source: Compiled by the authors.

Table 1.3 Types and examples of adaptation pathways

Type of Pathway	Explanation	Examples
Infrastructural	The assets, infrastructure, technologies, or "hardware" in place to ensure the delivery of services that could be disrupted by climate change (such as electricity or water)	Climate-proofing buildings; hardening seawalls and other seacoast structures against sea level rise; constructing reservoirs, irrigation systems, and wells to combat drought; maintenance of drainage systems; traditional rain and groundwater harvesting; and water storage and demand management
Institutional	The endurance of an institution or set of institutions, usually government ministries or departments, in charge of planning and community and infrastructural assets	Improving environmental monitoring, early warning systems for floods and other natural disasters, climate prediction and weather forecasting, vulnerability mapping, published evacuation routes, and public information on what to do during disasters
Community	The cohesion of communities and the livelihoods of the people that compose them	Developing agricultural systems and practices better suited to climate variability and change; land-use planning such as limiting development in low-lying coastal, wetland, and floodplain areas and other actions to limit precarious land uses; risk pooling at the regional or national scale; use of social care networks to reach vulnerable groups; and implementation of relocation systems for low-lying regions where storm surges may be severe
Ecosystem	The ability of ecosystems, habitats, and species to survive or even thrive in the face of changing climate	Transplanting species and establishing and maintaining gene banks to preserve biodiversity; and implementing mangrove conservation, restoration, and replanting to mitigate inundation related to rising sea levels

Source: Modified from Sovacool, B.K., D'Agostino, A.L., Meenawat, H., & Rawlani, A. (2012). Expert views of climate change adaptation in least developed Asia. *Journal of Environmental Management*, 97(30/April), 78–88.

"change location," for example, relocating people or activities away from harm, such as the establishment of flood evacuation centers, provision of boats for refugees, or providing climate migrants with new housing and land. A final is "restoration," restoring ecosystems or other infrastructure to their original conditions following climate-related damage. As Table 1.4 indicates, in the water and agriculture section of a survey of adaptation efforts in Bangladesh, the most common forms of adaptation center on modifying threats, changing locations, preventing losses, and sharing losses. Another related, yet distinct, way of categorizing adaptation is by the three notions of "protect," "retreat," or "accommodate;" or, decrease the probability of a climate impact occurring, reduce the impact by limiting its effects, or increase society's ability to cope with its consequences.[29]

2.2 Vulnerability and risk to disasters

Adaptation, in this way, centers on reducing vulnerability to disasters and other risks arising from climate change. The IPCC has defined vulnerability as "the predisposition of a person or group to be adversely affected...by the likelihood of severe alterations in the normal

Table 1.4 Example of water and agricultural adaptation efforts by type in Bangladesh

Impact/Event	Type of Adaptation	Example(s)
Flooding	Prevent effects	Flood control infrastructure
	Modify threat	Improvement in warning system
	Modify threat	Build flood tolerant infrastructure
	Share losses	Insurance
	Share losses	Credit relief
	Change use	Alternate cropping patterns
	Change location	Flood evacuation centers
	Change location	Taking refuge in boats
Drought	Modify threat	Increased irrigation
	Modify threat	Holding water in barrages during winter
	Modify threat	Augmentation of surface flow
	Prevent losses	Drought tolerant cropping practices
Salinity	Modify threat	Augmentation of surface flow
	Modify threat	Holding water in barrages for winter
	Prevent losses	Salinity tolerant crops
	Change use	Alternative land use

Source: Modified from Ahmed, A.U., Alam, M. and Rahman, A.A. (2011). Adaptation to climate change in Bangladesh: Future outlook. In S. Huq, Z. Karim, M. Asaduzzaman, & F. Mahtab (eds), *Vulnerability and Adaptation to Climate Change for Bangladesh* (pp. 125–143). London, UK: Kluwer Academic Publishers.

functioning of a community or society due to weather or climate events interacting with vulnerable social conditions."[30] The IPCC's fourth assessment report analogously defined climate vulnerability as "[...] a function of the character, magnitude, and rate of climate variation to which a system is exposed, its sensitivity, and its adaptive capacity." The report defined *sensitivity* as the degree to which a system is affected and *adaptive capacity* as the ability of a system to adjust to climate change. The evolution of the concept is visible in the fifth assessment report of the IPCC where the definition of vulnerability has broadened to include not only systems but institutions, humans, and other organisms that adjust to potential damage, that take advantage of opportunities, or that respond to consequences.[31]

These are verbose ways of saying that adaptation attempts to mitigate different types of disasters,[32] ranging from the "continuous," "recurrent," or "common" hazards that affect many people over many days to the "discrete," "singular," and "catastrophic" events which impact thousands or even millions of people, as shown in Table 1.5.[33] In addressing

Table 1.5 Continuum of vulnerability and disaster risk

	"Everyday" Disaster	"Small" Disaster	"Large" Disaster
Frequency of exposure	Continuous	Frequent or seasonal	Infrequent
Scale of impact	Individuals	Less than ten killed or 100 injured, no call for external assistance, no state of emergency	Ten or more killed, 100 or more injured, external assistance required or a state of emergency
Type of hazard	Environmental health hazards including air pollution and traffic accidents	Localized natural or technological hazards	Major natural or technological hazards
Form of risk management	Public health and critical infrastructure providers	Local emergency and recovery agencies	External emergency, recovery, and reconstruction agencies
Severity of risk	Low impact, high probability	Medium impact, medium probability	Large impact, low probability

Source: Modified from Pelling, M. (2011). The vulnerability of cities to disasters and climate change: A conceptual framework. In H.G. Brauch et al. (eds), *Coping With Global Environmental Change, Disasters and Security* (pp. 549_558). Berlin, Germany: Springer.

these disasters, vulnerability is not just exposure or sensitivity to them (the so-called external stresses) but also the capacity of an individual or community to be resilient to them, to evolve in order to accommodate the hazard or change positively to cope with it.[34]

Sometimes, different types of vulnerability are distinguished. O'Brien et al. differentiate between "outcome" and "contextual" definitions of vulnerability.[35] "Outcome" involves a linear assessment of biophysical impacts based on climate scenarios to identify adaptation options.[36] "Contextual," on the other hand, departs from identifying the capacity to respond to external stressors.[37] This definition seeks to recognize the multiple dimensions of socio-ecological factors. O'Brien et al. argue that these two framings of vulnerability in the literature complement each other, but are difficult to combine as they represent radically different understandings of what is at stake and what are the root causes.[38] The focus on projected biophysical impacts rather than on non-climate factors of vulnerability has been criticized by several scholars as risking being maladaptive and perpetuating vulnerability, as it does not provide the whole picture of what makes individuals or societies at risk to climate change.[39]

2.3 Resilience and adaptive capacity

Other terms that frequently recur in the climate adaptation literature are resilience and adaptive capacity. Resilience is slightly different to adaptive capacity. Resilience goes beyond adaptation to refer to the capacity of individuals or groups to implement many proactive adaptive actions over an extended period of time.[40] It is therefore similar and often synonymous with adaptive capacity, the ability of individuals and communities to adapt to adversity and stressful conditions, and reorganize into networks and institutions that learn and solve problems,[41] or what one study calls the "system's ability to prepare for, avoid, moderate, and recover from climate exposure and climate risk in general."[42] The framing of resilience or adaptive capacity is multidimensional, cutting across concepts from engineering, ecology, and sociology.[43]

Of course, all of these concepts – adaptation, vulnerability, adaptive capacity, and resilience – are interconnected and interrelated. The idea is that successful adaptation projects reduce vulnerability to climate hazards, strengthen resilience and the ability for social or ecological structures to absorb shocks, and bolster overall adaptive capacity to handle future threats.[44] Climate change adaptation serves as the gateway through which these positive outputs can be achieved, the idea being that each particular climatic threat can have its own set of adaptation mechanisms that improve overall resilience and adaptive capacity. Table 1.6 illustrates this interconnection in relation to a selection of climate threats from around the world.

A final term that sometimes circulates can be informally called "anti-adaptation" or "maladaptation." Early definitions referred to "those actions which tend to increase vulnerability to climate change. It is

Table 1.6 Interconnection of climate vulnerability, adaptation, resilience, and adaptive capacity

Vulnerability	**Adaptation Mechanism**	**Sources of Resilience and Adaptive Capacity**
Drought in Kenya and Tanzania	Switching occupation, selling assets, drought relief	Social networks, remittances
Drought in northeast Brazil	Private actions: livelihood diversification, risk management in agriculture, patron–client relationships Public actions: humanitarian relief, crop insurance, seed distribution, irrigation schemes	Lessons learned from past drought events, e.g., honed emergency relief mechanism, social networks, social security payments
Coral reef stress associated with physical damage, eutrophication, and fisheries decline in Tobago, West Indies	Development of community-based resource co-management, community monitoring of reef use, consensus building for future zoning and limitations on sewage disposal	High diversity in use between tourism and subsistence activities, heightened awareness of critical thresholds and well-defined user communities, learning through consensus building
Actual and potential disruption from hurricane in Cayman Islands, West Indies	Regulatory changes: enhanced building codes and zoning to increase waterfront setback; development of National Hurricane Plan Organization changes: creation of National Hurricane Committee and inclusion of diverse interests within it	Self-efficacy facilitated by high government revenue stability; recent experience of hurricanes (Hurricane Gilbert 1988, Mitch 1998, Michelle 2000, and Ivan 2004) promoted urgent learning from each experience accompanied by a willingness to learn from past mistakes; strong national and international support

Source: Modified from Nelson, D.R., Adger, W.N., & Brown, K. (2007). Adaptation to Environmental Change: Contributions of a Resilience Framework. *Annual Review of Environment and Resources*, 32, 395–419.

possible to make development or investment decisions while neglecting the actual or potential impacts of climate change. Such decisions are termed maladaptive."[45] The IPCC signaled a broader understanding of maladaptation in its 2001 report, which defined the term as: "Any changes in natural or human systems that inadvertently increase vulnerability to climate stimuli; an adaptation that does not succeed in reducing vulnerability but increases it instead."[46]

3 The case for adaptation

Now that we've spelled out what adaptation is and isn't, why exactly is it necessary for different groups, sectors and societies to cope with a changing climate? This section of the book touches upon five major justifications: efficacy compared to mitigation, precaution against tipping points, political feasibility, cost-effectiveness, and moral responsibility.

3.1 Efficacy compared to mitigation

Numerous catastrophic weather-related events over the last decades show that we need to better adapt societies to the effects of changing climate. Adaptation is all the more necessary given that humanity seems unlikely to mitigate greenhouse gas emissions in time to prevent major damage and disruption. From 1906 to 2005, the planet already warmed about 0.75 degrees Celsius and, depending on how societies develop over the coming decades, models predict global temperatures will increase by 2 to 6 degrees by the end of 2100.[47] Much of this global warming and climate change is already "locked-in" or "committed," meaning that even if all emissions were immediately halted – perfect and complete mitigation was miraculously achieved – warming would still continue for several decades as the oceans come into equilibrium with the atmosphere, a process known as "thermal inertia."[48]

This inertia could, under some scenarios, already result in a sea level rise greater than one meter, as Figure 1.1 depicts, making adaptation strategies of paramount salience. A 2014 *Nature* paper shows that the rate of global mean sea level rise was 1.2 millimeters per year from 1901 to 1990 (± 0.2 millimeters, 90 percent confidence interval). However, between 1993 and 2010 the annual mean sea level rise increased to 3.0 millimeters (± 0.7 millimeters), which is a much larger increase than previously predicted. Even if there is still significant uncertainty regarding its scale and scope, strategic adaptation needs to anticipate rising sea levels.[49] As Schneider put it succinctly, "adaptation is essential now to assist those who will likely be harmed by 'in the pipeline' climate change."[50]

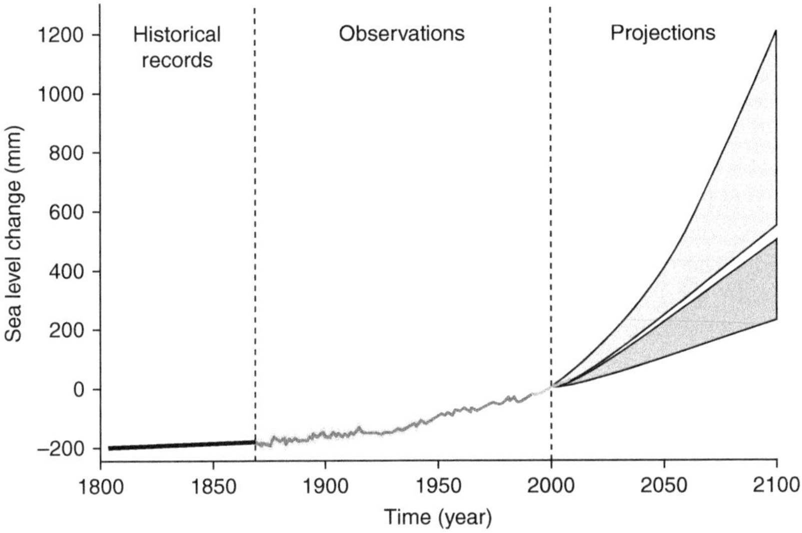

Figure 1.1 Historical and expected global sea levels, 1800–2100

Note: The thick black line represents the long-term sea level based on various observations for the nineteenth century. The grey line is based on tide gauge data. The light grey box is from satellite altimetry since 1993. The dark grey shaded region includes projections from coupled climate models.

Source: Cazenave, A. & Llovel, W. (2010). Contemporary sea level rise. *Annual Review of Marine Science*, 2, 145–173.

3.2 Precaution against tipping points

A related motivation for adaptation is that it acts as an insurance policy against unknown but potentially serious "tipping points." Climate-related tipping points are levers or scenarios where once a threshold is passed, "abrupt and rapid climate change" becomes inevitable.[51] In climate policy parlance tipping points are known as "bifurcations," special points in the climatic system where the deterministic part, the part responsible for its functioning, becomes irreparably altered.[52] Scientists are only just beginning to understand potential tipping points, but some of them could include:

- The melting of arctic summer sea ice, whose areal extent is affected by the temperature of the ocean;
- The melting of the Greenland Ice Sheet or West Antarctic Ice Sheet, whose ice volumes are affected by changes in air temperature;
- Disruptions to the El Nino Southern Oscillation, affected by thermo-haline depth;

- Intensified Indian summer monsoons and Sahara Sahel and West Africa Monsoons, affected by albedo, amount of vegetation, and precipitation;
- Accelerated deforestation of the Amazon rainforest and other Boreal rainforests, affected by precipitation and dry seasons; and
- The venting of methane in the East Siberian Arctic Shelf, affected by thawing permafrost.[53]

Under the most severe projections, the Arctic Ocean could be ice-free as early as 2037.[54] If the Greenland Ice Sheet melts, sea levels could rise a whopping six meters – enough to inundate almost all low lying island states as well as coastal areas from San Francisco and New York to Amsterdam and Tokyo.[55] The destabilization of ice shelves and the sudden and unexpected collapse of the West Antarctic ice sheet now have some scientists predicting "even greater likelihood of sea level rise in key regions.[56]"

When dealing with potential tipping points such as these, conventional notions of risk go out the window. Instead, analysts are forced into the domain of an "unholy trinity" of fat tails, tail dependence, and micro-correlations which mean the "uncertainty and risk profiles [concerning] climate tipping points are potentially loaded with catastrophic problems that defy the conventional bell-shaped probability curves."[57] In other words, most people are used to thinking of climate change as the turning of a dial, but a better metaphor for some of the impacts may be the flicking of a switch. Changes in nature have proven to be full of surprises, irreversible alterations, and non-linear interactions. While we might be uncertain about what limits are absolute and what the thresholds are, adaptation becomes an elemental option for better preparing and insulating society against what lies beyond potential tipping points.

3.3 Political feasibility

A third advantage is the relative political feasibility of adaptation policies and measures. Mitigation produces "public goods" and "global commons" problems, since a ton of carbon emitted anywhere in the world will be mixed into the atmosphere and contribute to climate change.[58] So as long as the positive effects for economy and health, for instance, are not taken into account, globally shared mitigation actions become a classic "tragedy of the commons" case where people must work together to stabilize anthropogenic interference with the climate system. So far, that sort of effective global collaboration has not occurred.[59] Adaptation, though, is different; it is not as prone to "free riders" and is something that all countries and communities can do for their own benefit. Dutch engineers raising dikes to adapt to rising sea levels benefits (arguably) all people in the Netherlands. In the words of one study, "it is useful to adapt even if nobody else does, but mitigation is meaningless unless it is as part of collective global effort."[60] Even if the statement is not entirely true, it points at the no-regrets potential of adaptation projects.

That said, well-designed adaptation interventions can still provide mitigation benefits. Some forms of adaptation can act as mitigation, like farming techniques that help store moisture better (enhancing adaptation) and also reduce the need for fossil-fuel intensive fertilizers (enhancing mitigation).[61] The afforestation of trees can act as a carbon sink (mitigation) but also can protect against flooding and provide income to communities (adaptation).[62] More efficient space cooling and heating can lower electricity consumption (a form of mitigation) but also can make cooling more affordable for lower income groups (adaptation); similarly, more efficient irrigation (a form of adaptation) can lessen the electricity needs associated with water pumping (a form of mitigation).[63] Restoring coastal wetlands can both increase carbon storage (mitigation) and create storm buffers against floods, fish nurseries, and new habitats for species (adaptation).[64] Reducing the exploration and drilling of offshore oil can prevent greenhouse gas emissions from fossil fuel combustion (mitigation) and diminish the risk of oil spills and consequent stress on ecosystems (adaptation).[65] Residential energy efficiency investments can not only reduce energy use (mitigation) but can also cut consumers' energy bills, translating into greater financial resilience to future shocks (adaptation).[66] Indeed, one assessment of the Global Environment Facility's Least Developed Countries Fund – the largest international scheme to promote investment in adaptation among more than 60 of the world's poorest countries – found that investment *mitigated* emissions as an unintentional, but positive byproduct, for a cost of less than $2 per ton, displacing carbon dioxide 2.35 to 500 times more cheaply than other mitigation options.[67]

3.4 Comparative cost-effectiveness

As a fourth plus, many investments in adaptation have positive cost-curves. That is, they make money and adapt to climate change at the same time, meaning they are "no-regrets" or "low-regrets" actions. In eight case studies, including drought in northern China and flooding in Florida, one study found that no- or low-regrets actions could address anywhere from 40 to 100 percent of the total expected loss from climate change.[68] Some of these options, such as improved drainage, sea barriers, and building codes, have payback ratios greater than 2 to 1 – it costs less than half as much to adapt now than to pay damages later. Similarly, researchers in Vietnam have calculated that investments in typhoon resistant housing pay for themselves in a matter of years, with most houses seeing internal rates of return ranging from 11 to 21 percent, depending on the severity of climate impacts.[69] The Asian Development Bank estimated that every $1 invested in adaptation in 2010 across Southeast Asia could yield as much as $40 in benefits by 2030.[70] Another study projected that every dollar invested in adaptation, globally, will yield $5 in benefits by 2030.[71]

3.5 Moral responsibility

Lastly, many experts argue in favor of adaptation on the grounds of fairness or morality. On a historical basis, the United States and Western Europe account for two-thirds of the primary buildup of carbon in the atmosphere, whereas the entire continent of Africa is responsible for a tiny sliver (just three percent) of global emissions.[72] However, depending on the methodology utilized, 75 to 80 percent of the costs of climate change will be borne by those in developing countries,[73] varying according to their national economic structures, geographies, and resilience.[74] Economically, more vulnerable countries are generally more dependent on climate-related sectors of economic activity such as agriculture, tourism, and forestry. Geographically, because of their general latitude, the net effect on ecosystems and communities for the bulk of developing countries will likely be negative (compared to a few high latitude regions, like Siberia, which could see small positive impacts). In terms of resilience, many vulnerable countries have higher rates of poverty and less institutional or technological capacity to cope with stress.

Indeed, looking at patterns of fatalities and economic damages from natural disasters, developing countries suffered 91 percent of fatalities and 51 percent of financial losses but accounted for just 12 percent of insured losses in 2007.[75] Over the period 1970 to 2008, the IPCC estimated that 95 percent of natural-disaster-related deaths occurred in developing countries.[76] Another global study distinguishing between "experienced" effects of climate change and "imposed" effects – essentially separating out those causing climate change and those experiencing it – concluded that people in rich countries impose 200–300 times more health damage from climate change on others from their historical GHG emissions than they experience themselves.[77] As one of the authors of the study, Berkeley professor Kirk Smith, mused, these aspects of climate change mean that it is "the most regressive tax in the world: the poorest pay for the actions of the rich."[78] As Abu Mostafa Kamal Uddin, former manager of the Bangladeshi government's climate change cell, convincingly argued, "During the global financial meltdown, trillions of dollars were mobilized to save the world's banks. What's wrong with helping the poor people of Bangladesh adapt to a situation we had nothing to do with creating?"[79]

To be fair, the blame for these disasters does not lay solely with climate change. There is to date little evidence that the surging disaster costs can be solely attributed to human caused changes in climate. Such conclusive signals may well be detectable in the future, but most likely not for several decades, if we look at the current empirical findings and scientific explanations.[80] One reason for this is obviously that so many societal factors influence damage costs and it is difficult to make statistical correlations on rare and unique events. In the language of IPCC's 2012 special report on extreme events and disasters, "there is *medium evidence* and *high agreement* that long-term trends in normalized losses have not been attributed to natural or anthropogenic climate change."[81] The IPCC concludes that most studies of disasters attribute surges in losses due to extreme weather events "to increasing exposure of people and assets in at-risk areas" as well as "to underlying societal trends – demographic, economic, political, and social – that shape vulnerability to impacts."[82] Even if disaster losses cannot be statistically attributed to climate change at present, severe consequences of extreme weather events are happening and with present trends, they will rise in the future unless societies around the world manage to drastically reduce their vulnerability.

Therefore, when viewed under the lens of justice and ethics, the activities of one group of persons and countries overusing the atmosphere as their carbon "dump" have caused and continued to injure a different, much larger group. This is a matter of "corrective" justice since one group has engaged in wrongfully injuring another group, meaning the guilty should desist from their harmful actions and also compensate their victims for damages.[83] Polluters and industrializing countries, the thinking goes, should be required to "clean up their own mess" and support others in adaptation. Part of the argument is historical, since industrializing countries have emitted the most into the atmosphere to reach current levels. Another part is an argument from ecology, since one could characterize the atmosphere as a sink with limited space. Rich countries have exhausted the capacity of this sink, denied other countries their shares, and are required to pay compensation for this overuse.[84] Such inequalities must be "reversed" by imposing extra burdens on those countries and peoples responsible for inflicting and producing those inequalities.[85] Adaptation offers us the ability to subsume this responsibility, since "sustainable adaptation can be considered a first step in guiding responses towards social justice and environmental integrity."[86]

4 Conceptual framework

With the basics of adaptation and the necessity of it now described, it's time to turn to the concepts and methods we employ in the pages to come. We draw from four separate disciplines – political economy, political ecology, justice theory, and critical development studies – which present some ideas that might be unfamiliar to most energy and climate researchers.

In its broadest sense, the term *political economy* deals with how government, or the "state," interacts with the private sector, or "the market."[87] Part of this involves the area of inquiry here, notably material provisioning, or how the political–economic system distributes material (and even immaterial) costs and benefits.[88] As Castree writes, "Political economy enquires into the origins, character and distribution of wealth. Its principal question about wealth is: 'who gets what, why and with what consequences?'"[89] Bohle et al. go on to explain:

> Each political economy is ... a product of fundamental class processes (the processes by which members of society produce necessary and surplus labour) and subsumed class processes (the distribution of surplus by the appropriators) ... Particular resource endowments and patterns of entitlements are always embedded in a macro-structure provided by political economy – that is to say the national regime of accumulation.[90]

Political economy therefore involves the study of struggle, or the processes by which some actors benefit from particular systems or processes at the exclusion of others.[91]

The closely-related field of *political ecology*, in its broadest sense, also focuses on the influence of power relations and structural inequalities, but with a closer link to the human processes which degrade the natural environment.[92] Biersack and Greenberg argue that political ecology refers to the "culture of production, distribution, and exchange" within the socio-environmental system.[93] Watts suggests that political ecology deals with "access and control over resources and their implications for environmental health and sustainable livelihoods."[94] Bryant and Bailey write that political ecologists generally accept the idea that "costs and benefits associated with environmental change are for the most part distributed among actors unequally," which serves to reinforce or reduce existing social and economic inequity.[95] Robbins adds that political ecology research "tends to reveal winners and losers, hidden costs, and the differential power that produces social and environmental outcomes."[96]

A third relevant, related domain of inquiry is often termed with various prefixes to *justice*: environmental, climate, or energy "justice." Environmental justice is concerned with the distribution of environmental hazards and access to all natural resources; it includes equal protection from burdens, meaningful involvement in decisions, and fair treatment in access to the benefits.[97,98,99,100] Walker defines environmental justice's two central issues as how some consume key environmental resources at the expense of others and how decision-making is unequally influenced.[101] Climate and energy justice are closely linked, and refer to the fairness, virtue, or equity dimensions of actions and decisions concerning greenhouse gases and energy production and use.[102,103,104,105] This set of literature revolves around the notion of fairness to present generations, given that some people have disproportionate access to the benefits of energy or a more resilient community, and fairness to future generations, given that we will leave them with the legacy of a polluted atmosphere and a potentially unstable climate.[106]

A final supporting school of thought comes from *critical development studies* or theory. Though vast, this type of scholarship generally critiques the global process of economic development. These theorists argue that "development" has only enabled advanced capitalist states, and recently major emerging economies, to define, know, and "meet" the needs of the "underdeveloped," essentially establishing a hierarchy of societies and cultures.[107] Under this view, development agencies have maintained a largely technocratic approach to intervention and assistance, relying on centrally conceived projects run by development experts accountable to narrowly defined, quantitative indicators.[108] A core conclusion in critical development studies is that such approaches need to be countered by participatory[109,110] or reflexive[111] modes of development practice, which place a greater emphasis on local knowledge, and uncover the political nature of development activities.[112]

There is an obvious overlap between these four disciplines. Castree argues that political ecology can be regarded as a subset of political economy:

> [Political ecology] is political economy (not "economics") in the sense that it sees the creation and distribution of wealth as a process involving both regulation by extra-economic institutions (notably the national state and its ancillary bodies) and social struggle – struggle between different actors over their size of the economic pie (e.g., capitalists and workers), and/or with those contesting the wider implications of economic activity on ostensibly non-economic grounds (such as "deep ecologists" protesting against airport expansions and road building programs). As this last bracketed example implies, political economy – even if its many practitioners have not always acknowledged it – is simultaneously and necessarily political ecology.[113]

Likewise, theorists have suggested that political ecology and energy and climate justice are intimately linked – with some scholars arguing that political ecology focuses on the "justice" issues of the "South" whereas energy or environmental justice focus on issues of the "North."[114] Forsyth argues that "political ecology represents an integration of environmental knowledge and social justice."[115] Many of the criticisms espoused by the critical development theorists – exclusion, inequality, homogenization – are also discussed in the justice literature, especially the parts of it dealing with remedies such as due process and procedural justice.[116]

The considerable similarity between these four domains of inquiry convinced us that their tenets can be synthesized, when applied to climate change adaptation, into three concerns:

- Costs, or how the hazards and externalities of adaptation projects are imposed on communities unequally, often the poor, disempowered, or marginalized;
- Benefits, or how access to a particular adaptation project might be biased or unevenly distributed;
- Procedures, or how adaptation projects might proceed with exclusionary forms of decision-making rules and procedures that lack due process for involvement and representation.

With an appreciation for looking at how the costs, benefits, and procedures of adaptation might play out on the international stage, we propose the following four processes summarized by Table 1.7. The remainder of this section outlines each in turn.

4.1 Enclosure

The process of enclosure refers to when an adaptation project transfers a public or social asset into private hands, or expands the role and authority of a private actor into a formerly public sphere. It relates in part to how private institutions, especially corporate actors, intensify their efforts to penetrate into more isolated or peripheral areas from which they can derive revenue.[117] It can involve the incorporation of adaptation projects into the act of "territorial accumulation" and the "continuous self-expansion of capital within the global system."[118] In this way, adaptation can expand the reach of capital as it "stretches" the reach of the market to encompass remote areas, and "deepens" capitalism by allowing it to penetrate into the provision of yet more goods and services.[119] As Harvey writes, "the corporatization and privatization of hitherto public assets…constitute a new wave of 'enclosing the commons.'"[120] In short, adaptation projects become "enclosed" as part

Table 1.7 Summary of the processes and sub-processes of enclosure, exclusion, encroachment, and entrenchment

Process	Description	Sub-processes	Disciplines
Enclosure	Capturing resources or authority	Territorial accumulation, privatization, market stretching, parallel bureaucratization, land grabbing	Political geography, environmental geography, public policy, public administration, neo-Marxism
Exclusion	Marginalizing stakeholders	Dispossession, accumulation by dispossession, tyranny	Political ecology, climate policy, development studies
Encroachment	Damaging the environment	Commodification, subordination, forum shopping	Ecology, environmental science, biodiversity conservation, consumption studies
Entrenchment	Worsening social inequality	Comparative advantage, elite capture	Welfare economics, environmental justice, gender studies, political economy

Source: Authors.

of the strategy of capitalist accumulation.[121] The "enclosure" of climate change provides "another moment for the rapid creation and capture of value" by the global marketplace and those it serves.[122]

These types of enclosure can have many causal mechanisms.[123] In some cases, climate policies can convince corporate actors to invest in solutions that are land-intensive, such as genetically engineered crops or the growing of feedstocks, and tend to displace people. In others, land and natural capital might be directly appropriated for environmental ends, such as circumventing restrictions on logging in an old-growth tropical forest. In still others, national or corporate "land deals" may legally set aside land for other uses such as economic development or the creation of jobs.

Enclosure need not only occur with physical resources such as money or land – it can also relate to immaterial resources such as power and sovereignty. Enclosure in this way can refer to when an adaptation project seeks to work "around" rather than "with" a particular state structure, creating its own "parallel bureaucracy"[124] or "node of power" beyond the state.[125] It encloses upon the authority or expertise usually afforded to state or public actors.

This type of enclosure can occur when private actors subsume or privatize – enclose – roles that previously belonged to public and state actors. In Honduras, for instance, disaster recovery efforts in 1998, after Hurricane Mitch, enabled private actors to enclose upon activities formerly in the public sphere. Eight private organizations, mostly transnational Nongovernmental organizations (NGOs), resettled survivors in the Amarateca Valley and founded new communities – taking responsibility for water works, road construction, schools, housing, and other infrastructure – without government oversight.[126] This made the communities entirely dependent on NGOs for services provision. In a time when United Nations climate multilateralism is questioned, non-state or network climate governance has been thriving and new initiatives have steadily gained speed.[127] In the last decade, such climate governance has seen an increasing number of non-state actors assuming or being delegated governance functions.[128] This certainly goes for adaptation measures. Non-state actors are increasingly viewed as possible "agents of change" not only in implementing projects, providing resources, and raising awareness, but also in governing, for example, through public–private partnerships on adaptation.

In other situations, disaster recovery efforts can take advantage of property destruction to redistrict or redesign urban areas in elites' favor. The best known example might be the 1906 earthquake that ruined a large part of San Francisco in the United States. There, city leaders were successfully able to move Chinatown from its earlier, central and geographically favorable location to a more marginal neighborhood upon rebuilding.[129]

4.2 Exclusion

The process of exclusion often occurs in tandem with enclosure,[130] and it refers to when an adaptation project excludes or displaces a particular group of stakeholders or limits access to resources related to due process, fairness, and procedural justice. The process of exclusion enables resources to be appropriated or consolidated by state authorities, private firms, or social elites.[131] In many cases, relevant community-based organizations or individuals are excluded from the decision-making process. For instance, Satterthwaite, Dodman, and Bicknell argue that "the relevance of community-based adaptation is also underplayed in that the policies and practices of governments and international agencies do not recognize the capacity of community-based organizations to contribute to adaptation."[132] Exclusion can occur unintentionally, but it can also be a strategy of containment, a way to prevent and manage other actors from interfering with one's interests.[133] The process of exclusion, ironically, can also create dependence, making displaced communities dependent on those that exclude them as relations of production and consumption become cemented in ways that perpetuate their subordinate status.[134]

Adaptation projects can, under certain circumstances, pass on climate vulnerabilities through this exclusion. One example would be a situation where a project enables a large property owner to lease out small parcels of land for sharecropping by poor farmers. If a flood comes or crops fail, the tenants, rather than the landowner, suffer the most, because the risk has been "socialized" to the tenants even as they are excluded from the project's direct benefits.[135]

The process of "exclusion" is close to the notions of "dispossession" and "accumulation by dispossession" affiliated with capitalism, or that of "tyranny" affiliated with unfair decision-making processes. "Dispossession" traces its roots all the way back to Karl Marx, who held that the capitalist system is constantly striving for profits and capital accumulation in a competitive market economy so that labor becomes "dispossessed" and treated as a thing, a commodity, subject to the same pricing mechanisms.[136] Harvey has drawn from this classical concept to create a modern idea of "accumulation by dispossession," defined as the "centralization of wealth and power in the hands of a few by dispossessing the public of their wealth or land."[137] Accumulation by dispossession can take a variety of forms, including the privatization of land and forcible relocation of people residing there, the establishment of property rights or suppression of rights to the commons, and the process of appropriating assets such as natural resources or land.[138] Cooke and Kothari frame their investigation of exclusion with the language of "tyranny," arguing that decision-making processes themselves can become tyrannical and exclusionary.[139] One way this tyrannical exclusion can occur is when multilateral or international agencies and funders dominate discussions and decision-making about adaptation. Non-state actors can act on delegated authority from states. They can serve to provide accountability and legitimacy to the process as well as be instrumental for implementation, but the process also creates opportunity for some groups to strengthen their own interests.

4.3 Encroachment

The process of encroachment refers to when adaptation projects degrade the environment, interfere with ecosystem services provision, or intrude upon biodiversity conservation zones such as protected areas and national parks. Though research on how the process of biodiversity conservation itself can encroach upon indigenous lands or promote capitalist agendas is well documented,[140] less attention has focused on the other direction, namely how hegemonic adaptation projects can encroach upon ecosystems or create their own type of environmental degradation.

Adaptation can, because it is primarily concerned with building human resilience, undermine the conservation of biodiversity. Turner et al. write that "the natural systems upon which people depend, already under direct assault from climate change, are further threatened by how we respond to climate change" and that "human adaptation to climate change will have far-reaching effects, ricocheting between different biomes and affecting biodiversity through globally connected economies and markets...biodiversity is likely to be undermined by a broad and complex array of human responses to climate change."[141] As an example, reconstruction after the Indian Ocean tsunami in 2004 damaged ecosystems in multiple ways, such as sand and gravel extraction from river channels and the rapid siting of new housing without regard to environmental sensitivity.

Adaptation efforts can also sometimes lead to increased greenhouse gas emissions – encroaching upon (reducing the efficacy of) efforts to mitigate carbon dioxide. For instance, a community adapting by using more refrigeration or air conditioning in response to hot temperatures would also hurt mitigation efforts aimed at decreasing emissions from electricity consumption.[142] Efforts at reducing disease vectors such as malaria (enhancing adaptation) might require significant energy use, and associated wetlands conversion might increase greenhouse emissions.[143] Reducing the spread of water borne disease and providing potable water can require more energy use for water treatment.[144] The construction of desalination plants or interbasin water transfers can help offset salt-water intrusion and shortages of water but only at the expense of higher ongoing energy consumption to operate the plants or pump, store, and transfer water.[145] Relocating critical infrastructure outside of floodplains can enhance adaptation but requires increased emissions due to the rebuilding of structures; it also likely increases sprawl and associated emissions from transportation.[146] Farmers can adapt to climate change by increasing their use of nitrogen fertilizers to offset yield losses but only at the expense of higher emissions from the agricultural sector.[147] In these instances, pursuing one form of adaptation can actually conflict with other forms of mitigation, worsening climate change in the end.

Encroachment need not always be so direct. It can operate as part of a sub-process of commodification, one that transforms biodiversity and other protected assets into "productive" uses that we can utilize

Table 1.8 "Soft" and "hard" pathways for climate change adaptation

Soft Adaptation Path	**Hard Adaptation Path**
Promotes forms of natural infrastructure or natural capital, such as ecosystems and forests, together with low-impact technology	Promotes predominantly human-built infrastructure
Empowers local communities, and builds institutional capacity and community assets, puts "community wishes" first	Involves large-scale disturbances to local communities and/or ecosystems, puts "adaptation goals" first
Relies on simple and modular technologies, relatively easy to understand, that do not require large outlays of capital or human resources	Relies on complex and capital-intensive technologies
Prefers technologies and/or processes generally owned by local people	Prefers technologies and/or processes are generally owned by private firms or the government
Retains the ability to respond to alterations in climate change projections	Lacks flexibility and adaptability to sudden changes in projections of climate change

Source: Modified from Sovacool, B.K. (2011). Conceptualizing hard and soft paths for climate change adaptation. *Climate Policy*, 11(4), 1177–1183.

to combat climate change or turn a profit.[148] It can also operate as a sub-process of subordinating other policy or societal goals – such as protecting the sovereignty of indigenous peoples, or the wishes of rural communities – to adaptation goals, operating along a "hard" path shown in Table 1.8 rather than a more adaptive, decentralized "soft" path.[149] Encroachment can lastly occur as part of a sub-process of "forum shopping" where savvy investors or firms look for areas amenable to encroachment to keep costs low and shift externalities to publics with little protection from them.[150]

4.4 Entrenchment

The process of entrenchment refers to when an adaptation project aggravates the disempowerment of women or minorities. It "entrenches" inequality of groups, most notably by interfering with egalitarian systems of distribution, or by further concentrating wealth within a community or transferring risk.

Economic entrenchment occurs because it is typically wealthier households that possess the requisite assets to maintain resilience in the face of climate change – or in extreme situations, take advantage of vulnerable groups. It is these more powerful actors that can utilize cash or other productive assets to rebuild, rent buildings, purchase equipment, diversify livelihood opportunities or retrieve information that they can deploy to broaden and extend their influence. Poorer actors, by contrast, cannot generally afford to undertake these activities.[151] Holling, Gunderson, and Ludwig caution that adaptation resources could also be "re-appropriated by powerful minorities able to influence public policy in ways that benefit them," leading to a situation in which "inappropriate measures such as perverse subsidies are implemented that deplete resources and create inefficiencies" and encourage "rent seeking by powerful minorities."[152] Entrenchment, in other words, masks and maintains, and at times worsens, existing power relationships at the international, national, local or household level.[153]

Sub-processes closely linked to entrenchment include "comparative advantage" and "elite capture." Comparative advantage refers to the way in which adaptation projects can disrupt the dissemination of products and services that result in uneven distribution of benefits. Adaptation spurs transformations at local, national, and even global scales that comparatively benefit some and hurt others, enabling some to profit from climate change and gain comparative advantage.[154] Elite capture refers to the phenomenon whereby resources appropriated to the masses or poorer stakeholders become captured or usurped by politically or economically powerful groups at the expense of less influential and more peripheral groups.[155]

A final element of entrenchment is the transmission of risk from one vulnerable population to another, or from a wealthy population to a less resilient one. Structural measures such as dikes or seawalls, for instance, provide physical protection from floods or storms only by shifting those risks elsewhere, in the case of Bangladesh to those living outside or around the protective area. As one study warned, with these types of adaptation projects, "some areas have been made flood free, but the problem is transferred to somewhere else."[156] Sociologists Robert D. Bullard and Beverly Wright add that "disaster mitigation and investments provide location-specific benefits, restricted to populations that live or own assets in the protected areas. Thus ... some members are excluded from the benefits of these investments."[157] Adaptation projects therefore become an elemental part of "risk transference" between communities, which can introduce or heighten inequality.

5 Research methods

To collect data for our book and conduct our analysis, we decided to follow a comparative case study approach. We hold, as Stern has argued, that "nothing advances theory better than tackling a practical problem by integrating different perspectives."[158] We also maintain that such theoretical exploration can have a dual impact on both practice and theory. As sociologist Robert K. Merton noted almost 40 years ago, "sociological theory and research not only helps us to identify and to understand what goes on when disaster strikes but also, conversely, the investigation of these phenomena can extend sociological theories of human behavior and social organization."[159]

We decided to focus our inquiry around four different case studies using an adaptation project, or suite of projects, as our unit of analysis. As Table 1.9 reveals, these four cases involved different climate risks and adaptive technologies, time periods, regulatory regimes, market economies, and sectors. We selected Bangladesh because it is an emerging, lower-middle-income economy relying on coastal defenses, community livelihood diversification, and local coping strategies to deal with flooding. We selected the Netherlands because it is a socially democratic, fully industrialized European state that has invested heavily in dikes and storm surge barriers to protect its population from rising sea levels. We selected the United States because it is a corporatist democratic, post-industrialized society, home to the world's largest economy that still

Table 1.9 Summary of four political economy of adaptation case studies

Case Study	Time Period	Country	Type of Economy	Form of Governance	Predominant Climate Risk	Form(s) of Adaptation
Bangladeshi National Adaptation Program of Action (NAPA)	2005–present	Bangladesh	Lower middle income	Parliamentary political system	Flooding	Polders, diversification of employment (shrimp farming), coastal afforestation, household coping strategies
Dutch Delta Works	1958–1997	Netherlands	High income	Parliamentary political system	Sea level rise	Storm surge barriers, canals, and dikes
Hurricane Katrina reconstruction	2005–2008	United States	High income	Representative democracy	Hurricanes	Disaster recovery
United Nations Framework Convention on Climate Change	1992–present	195 countries	–	International treaty	Various	Various

Source: Authors.

managed to bungle recovery after a severe hurricane. We selected the UNFCCC because it is the premier piece of climate governance architecture (the main international governance arrangement responsible for climate change) regulating and affecting the transfer of adaptation practices and technologies between countries. Note that our cases therefore at times involve specific adaptation projects as well as coping strategies or adaptation processes.

We decided to rely on the presentation of qualitative case studies for numerous reasons. In developing case studies, we endeavored to provide what methodological theorists Alexander George and Andrew Bennet call a "detailed examination of an aspect of a historical episode to develop or test historical explanations that may be generalizable to other events."[160] Rather than utilizing laboratory samples or statistical analysis to examine variables, case study methods involve an in-depth, longitudinal assessment of a single instance or group of instances: a case or cases.[161] Put another way, the case study method is an investigation of a contemporary phenomenon within its real-life context to explore causation in order to find underlying principles. As Yin points out, case study methodologies allow researchers to delve deeper into a given phenomenon and develop a richer, more descriptive understanding of the subject under study.[162] The intention is to foster a more accurate understanding of a complex issue and provide detailed, contextual insight for students or policymakers and even analysts.[163]

After all, we hold that social science research is primarily the study of human affairs, and that only context-dependent knowledge really exists in this realm. Such knowledge is always related to specific instances which case studies are apt to capture. Elucidating the commonalities, and differences, between these cases is the key to student learning, building expertise, making comparisons, and enhancing critical thinking.[164] Case studies can also show policymakers successful models upon which they can base future projects. Case studies have the "irreducible quality" of being "thick," that is, full of rich detail, or "hard to summarize."[165] Only through an intimate connection with experience and reality can readers develop testable, relevant, and valid knowledge and information.

That said, our methodological approach does have some weaknesses. Case study work can produce "thick" descriptions but also "messy" ones, since it is hard to filter all of the qualitative data gleaned.[166] Furthermore, our case studies all relate to our dependent variable,[167] meaning that they all demonstrate negative examples of the four Es. To truly test to what extent adaptation programs are negatively impacted by enclosure, exclusion, encroachment, and entrenchment, we would also need to examine the potentially positive outcomes of the adaptation efforts affected by the four Es.

Ultimately, however, we believe that the strengths of our methodology far surpass its weaknesses. Selecting cases related to a dependent variable, or cases that are all similar, can have the advantage of great depth.[168] Similarity can also matter if the phenomenon appears to occur across a complex number of cases that differ based on time, geography, and context. Here, with our case studies, it is the commonalities that are striking. The elements of the four Es appear despite the rich variety of our examples.

6 Preview of chapters

In sum, drawn from our case study methodology, we believe that these four processes of enclosure, exclusion, encroachment, and entrenchment are significant in many respects. Their negative consequences are familiar but grim, including: "increases in socially produced scarcity, growing inequality, and often accelerated depletion or degradation of the very resources market mechanisms were supposed to protect."[169] Once these processes begin, they can become difficult to reverse, as adaptation becomes "central to the commodification of various parts of the landscape, laying the bases for new rounds of landscape transformation as businesses seek to gain access to these commodities."[170] Because scarcity tends to lead to increases in the exchange value of a commodity in the marketplace, the destruction of the environment and commodification of people and ecosystems that result from the cooptation of adaptation can "inevitably increase profit opportunities for some at the expense of others."[171]

To support these arguments with real-world examples, the remainder of the book proceeds as follows. Chapter 2 looks at how projects and coping mechanisms being implemented under Bangladesh's NAPA have increased the displacement of indigenous people. Bangladesh is one of the nations most vulnerable to the impacts of climate change, yet this chapter shows how enclosure, exclusion, encroachment, and entrenchment impede the vitality of its adaptation efforts. Climate change policies have enabled elites to capture land through public servants, the military, and even gangs carrying bamboo sticks. Exclusionary forms of adaptation planning exist at both national and local scales. Climate protection measures have encroached upon village property, *char* (public) land, forests, and farms. Most egregiously, community coping strategies for climate change have entrenched class and ethnic hierarchies, ultimately trapping the poor, powerless, and displaced into a feudal patronage system that only aggravates human insecurity and intensifies violent conflict.

Chapter 3 investigates how the Delta Works Project in the Netherlands has eroded biodiversity and ecosystem integrity. A central part of the Works, the Eastern Scheldt Storm Surge Barrier (ESSSB), or *Oosterscheldekering*, is a nine-kilometer-long, $4.7 billion protective seawall in southwestern Netherlands. An engineering marvel conceived and built under the direction of the *Rijkswaterstaat* (the national public works organization), it was intended to reduce flooding risks to 1 in 4,000 years. As this chapter demonstrates, the Delta Works enabled the *Rijkswaterstaat* to enclose upon the autonomy and sovereignty of local water bodies, and excluded fishers, recreational boat owners, local politicians, and some environmentalists from decision-making. It encroached significantly on the unique wetlands and salt marshes of the Eastern Scheldt estuary and led to an entrenchment of flooding risk in other parts of the Netherlands.

Chapter 4 reveals how reconstruction efforts in the United States after Hurricane Katrina worsened patterns of racism, classism, and sexism. Over the course of August and September 2005, Hurricane Katrina became the costliest natural disaster ever faced by the United States, one that flooded 80 percent of New Orleans and 90,000 square miles of land across Louisiana, Florida, Mississippi, and Texas. This chapter shows how the political economy of recovery efforts played a significant role in exacerbating the storm's $80 to $400 billion price tag, hundreds of deaths, and millions of evacuees. The Katrina recovery process expanded corporate control and profiteering over previously public services. Evacuees found themselves excluded from transportation, housing, reconstruction jobs, decision-making venues, and insurance relief. Private actors used Katrina reconstruction to relax environmental protections and aggressively promote infrastructure repair in ways that damaged the environment. Lastly, Katrina reconstruction further entrenched class-, race-, and gender-based inequality as well as unequal levee protection.

Chapter 5 analyzes unequal access to knowledge and technology within the UNFCCC. It is the lynchpin of an entire international regime dedicated to addressing global warming. Despite good intentions, discussions and negotiations within the Conference of the Parties are rarely representative, running the evident risk of steering international negotiations in the direction of vested interests. Related adaptation projects have resulted in deforestation or changes in land use. Further, there is a discernible lack of fair (and promised) funding for adaptation as well as limited technology transfer via the UNFCCC, which keeps the South locked – entrenched – in a relative state of technological inferiority and macroeconomic inequality.

Chapter 6 argues that many of the unjust and inequitable aspects of the climate change adaptation projects examined in this book – enclosure, exclusion, encroachment, and entrenchment – can be managed or at least minimized by a set of five best practices matched in scale and scope with five design principles. The chapter first proposes five design principles centered on: distributive justice and equity, procedural justice and organizational multiplicity, information and transparency, accountability and responsibility, and adaptability and arbitration. These design principles lead to five policy mechanisms, or best practices: taxing adaptation's externalities; critically involving stakeholders; seeking freely given, prior, informed consent; implementing bonds and instruments of liability to offset damages and creating adaptation inspection panels; and forming an international tribunal to handle redress.

Chapter 7 presents five conclusions and policy implications resulting from the analysis presented earlier in the book. Our conclusions are that (1) the processes of enclosure, exclusion, encroachment, and entrenchment can subvert the intended goals and desirable attributes of adaptation projects; (2) conflict and struggle are inescapable in adaptation projects; (3) adaptation projects are enmeshed in complex global relationships and phenomena; (4) a holistic array of five policy options needs to be implemented to hold the political economy processes of adaptation in check; and (5) future research is needed to strengthen our understanding of both adaptation and its political economy. Because of the underlying political economy of adaptation, analysts, practitioners, and planners need to rethink their fundamental assumptions about when adaptation projects are warranted, and what criteria ought to guide assessment of their successes or failures.

2

Bamboo Thumping Bandits: The Political Economy of Climate Adaptation in Bangladesh

1 Introduction

Bangladesh contributes little to global greenhouse gas emissions, yet it is one of the countries most vulnerable to climate change. Because it sits at the intersection of three major river basins, illustrated in Figure 2.1, and features flat deltaic topography with low elevation, it is prone to a multitude of climate-related events such as floods, droughts, tropical cyclones and storm surges. Fifteen percent of its 162 million people live within one-meter elevation from high tide,[1] and annual floods inundate between 20 and 70 percent of the country's landmass each year.[2] Bangladesh has high population density and rates of poverty. It is the seventh most populous country in the world, with a density greater than one thousand persons per square kilometer.[3] Yearly per capita income ranges from $400 to $1,700 (depending upon what counts and whether purchasing power parity is considered).[4] Bangladesh also has extreme climate variability, naturally alternating between seasons of monsoon and winter drought, and the nation is dependent upon crop agriculture, which is highly sensitive to changes in climate.[5] Two-thirds of Bangladesh's critical infrastructure sits five meters or less above sea level and Bangladesh is the most vulnerable country in the world to tropical cyclones and the sixth most vulnerable to floods.[6]

Thus, climate change adaptation has been recognized as necessary to the political and economic survival of the country.[7] However, after previewing Bangladeshi vulnerability to climate change, this

Figure 2.1 The river systems and geographic location of Bangladesh

Source: Modified from Mirza, M.M.Q. (2002). Global warming and changes in the probability of occurrence of floods in Bangladesh and implications. *Global Environmental Change*, 12(2), 128.

chapter demonstrates the detrimental presence of enclosure, exclusion, encroachment, and entrenchment in climate change adaptation efforts. Climate change policies have enabled rural and urban elites to capture land. Exclusionary forms of adaptation planning and implementation exist at national and local scales. Climate protection measures have led to encroachment upon the public commons. Finally, community coping strategies for climate change have, at times, entrenched existing class and ethnic hierarchies that trap the poor, powerless, and displaced into a nefarious system of patronage that only accelerates human insecurity and perpetuates violent conflict.

Before we delve into our extensive critique of Bangladeshi adaptation, however, we need to contextualize our findings. Though the situation in Bangladesh looks bad, the fact that we have demonstrated the existence of enclosure, exclusion, encroachment, and entrenchment in some Bangladeshi adaptation measures does not mean that they are always present or even frequently present. There are many adaptation projects that seem to be producing a net social benefit despite the complex Bangladeshi political economy surrounding them. To highlight two examples: one small-scale project in Bangladesh has reduced the vulnerability of women in *char* areas and minimized gender disparities in coping strategies to climate change.[8] That project saw female heads of household increase their number of goats and poultry; wages between men and women equalize; and women join in other income-generating activities such as rice processing or rickshaw driving.[9] Interestingly, an independent evaluation of the program found that the longer participants were involved in NGO and group activities, the greater the positive impact – that is, benefits multiplied the longer the program operated.[10] A second project demonstrated the integration of climate-proofing infrastructure – in this instance, mangrove afforestation – *with* community income generation, specifically the planting of crops and fishing, in what has been termed the "Triple F" model of resilience.[11] The coastal communities most vulnerable to rising sea levels – the places where mangroves need to be planted and forests replenished – are also those where farming and forestry are the primary sources of income. The "FFF" model attempts to maintain community livelihood and adapt to climate change simultaneously by integrating aquaculture and food production within reforested and afforested plantations.[12] So, not every adaptation project need perpetuate inequality, exclude others, or enclose and encroach upon people's property or livelihood. Unfortunately, some do.

2 Vulnerabilities to climate change in Bangladesh

We selected Bangladesh for our first case study due to its extreme vulnerability to climate-related impacts. Most of Bangladesh lies in the delta of three of the largest rivers in the world: the Ganges, the Brahmaputra, and the Meghna, or GBM. These rivers have a combined peak discharge of 180,000 cubic meters per second during the flood season, the second highest in the world after the Amazon, and carry about two billion tons of sediment each year. Bangladesh is at risk not only to flooding and tidal inundation on the coasts, but also advanced melting of the Indian and Nepali Himalayan glaciers. This effectively means that the country is hit on "both geographic sides" as well as during "both seasons;" climate change risks disrupting natural cycles of rainfall and snowpack on the Tibetan Plateau, which feeds Bangladesh's major rivers, and also increasing flooding, saltwater intrusion, and storm surges on the coastal belt.[13] Climate change is also creating excess rain during the monsoon season, and inducing a shortage of it during the winter drought.[14]

Further compounding matters, the topography of the country is low and flat. Two-thirds of its critical infrastructure is less than five meters above sea level and is therefore susceptible to naturally occurring river and rainwater flooding and, in lower lying coastal areas, to tidal flooding during storms. Indeed, one study proved that Bangladesh is perpetually at risk to four distinct types of flooding:[15] *flash floods* which occur from rapid runoff of rainwater in hilly regions; *riverine floods* which result when the GBM rivers or their tributaries, which tend to rise and fall slowly over a ten to twenty day cycle, simultaneously reach their peaks; *rain floods* due to high intensity local rainfall during the monsoon season; and *storm surge floods* caused by tropical storms and cyclones which affect tidal flats and low-lying islands.

Once every four to five years, severe flooding wreaks substantial damage to infrastructure, housing, agriculture, and livelihood. Table 2.1 shows serious floods in Bangladesh over a 25-year period and their impacts. A severe tropical cyclone also hits Bangladesh, on average, every three years resulting in extensive damage to houses, livestock, and human health.[16] One cyclone in November 1970 produced winds as strong as 220 kilometers per hour and a storm surge nine meters high, resulting in 300,000[17] to 500,000 deaths.[18] Another cyclone in April 1991 generated a storm surge eight meters high that displaced 11 million people and caused 138,000 deaths.[19]

Indeed, Bangladesh has been hit by 154 cyclones from 1877 to 1995, a rate of more than one cyclone per year, and also subject to 174 separate

Table 2.1 Impacts of major floods in Bangladesh, 1984–2007

Year	Land Inundated (square kilometers)	Fatalities	Refugees/ Homeless	Damage (in Millions of $ 2007)
1984	50,000	445	N/A	378
1987	50,000	2,055	N/A	1,000
1988	85,000	6,500	45 million	1,200
1998	100,000	1,100	30 million	2,800
2004	60,000	700	15 million	6,600
2007	62,300	1,100	1.1 million	1,100

Source: Adapted from Government of Bangladesh (2009). *Bangladesh Climate Change Strategy and Action Plan*. Dhaka, Bangladesh: Ministry of Environment and Forests (MOEF).

natural disasters from 1974 to 2003.[20] Another more recent analysis notes that Bangladesh experienced 21 above normal floods, four exceptional floods, and two catastrophic floods from 1954 to 2010.[21] One flood in 1988 reduced agricultural production by 45 percent, and a 1998 flood resulted in the loss of more than two million hectares of arable land used for rice cultivation.[22] A more recent severe flood, in 2007, inundated 42 percent of the country's land area (62,300 square kilometers), caused 1,110 deaths, submerged 2.1 million hectares of cropland, destroyed 85,000 homes, damaged 31,000 kilometers of roads, affected 14 million people, and induced $1.1 billion in damages.[23] To put the damage into perspective, $1.2 billion is equal to all public debt listed by the government in 2010.[24] Figure 2.2 illustrates how between 1954 and 2009, six major floods have inundated more than one-third of the country's landmass. So, even without future impacts caused by human interference with the climate system, Bangladesh is in dire need of adaptation measures.

Disturbingly, such floods and natural disasters are projected to get worse over the next few decades. A synthesis of 16 General Circulation Models and three emissions scenarios in the Intergovernmental Panel on Climate Change's AR4 report expects Bangladesh to see temperature increases of 1 to 3 degrees Celsius by 2050.[25] Basically, this warming will create problems associated with water: too much of it during the monsoon seasons, and too little of it during the winter. Temperature increases will likely raise sea levels; increase river water levels, water logging, erosion, and flooding during the monsoon season; and exacerbate saltwater intrusion and shortages of water for irrigation and agriculture during the winter.[26] Agrawala et al. anticipate four primary negative changes in climate and precipitation: accelerated glacier

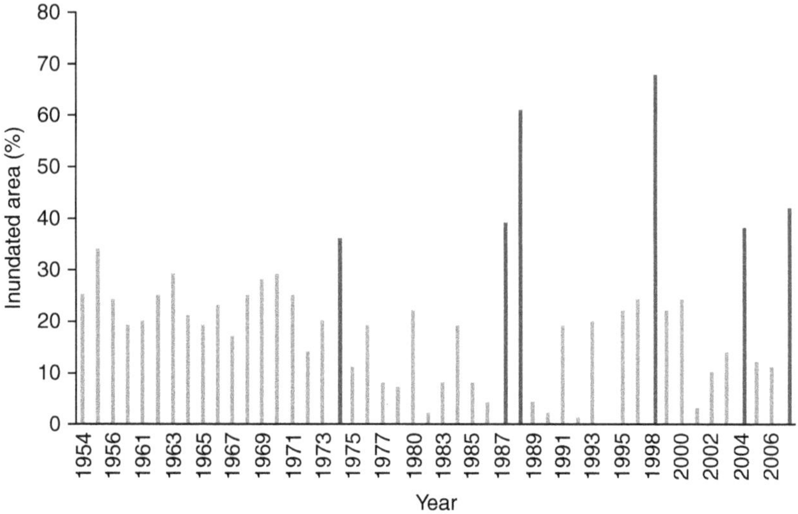

Figure 2.2 Extent of above normal flooding in Bangladesh, 1950–2009

Source: Modified from Dasgupta, S., et al. (2010, November). Climate proofing infrastructure in Bangladesh: The incremental cost of limiting future inland monsoon flood damage. Policy Research Working Paper 5469, (p. 29). Washington, DC: World Bank.

melting from increased runoff from the neighboring Himalayas, increased rainfall during the monsoon season, sea level rise leading to flooding under ambient conditions and severe flooding during storm conditions, and increased frequency and intensity of cyclones.[27] Every area in Bangladesh is prone to at least one out of the four of these negative changes.

Adding to a pessimistic outlook, earthen embankments constructed by the Bangladesh Water Development Board are subject to erosion, and with a 45 cm rise of sea level, experts expect 15 percent of the land in Bangladesh to be inundated by the year 2050, resulting in more than 25 million climate refugees from the coastal districts. Especially vulnerable in these locations would be *char* dwellers and women. *Char* dwellers are vulnerable because they live in constantly changing islands, or chars, in the coastal district, but build *kacha* homes made of *muli* bamboo, mud, and tin roofs that can be easily swept away in floods. Women are at risk because they contribute a disproportionate amount of their time rebuilding homes and caring for family members.[28] The World Bank has warned that a 25 cm increase in sea level, quite likely given recent projections, would force Bangladesh to lose 6,300 square kilometers (four percent of its land), make the country prone to a 1991 level cyclone of ten percent greater intensity, provoke monsoonal floods, increase crop losses, inundate 40 percent of the Sundarbans, and increase the salinity of soil and water.[29] Belt reports that rises in sea level could place as many as 30 million people along the southern coastal belt of Bangladesh at risk of displacement,[30] and a recent study of 136 global cities concluded that the two cities likely to witness the greatest proportional increase in people exposed to climate extremes by the year 2017 were Dhaka and Chittagong, both in Bangladesh.[31] Agrawala et al. similarly projected that a one-meter rise in sea level would threaten 18 percent of Bangladesh's population, inundate one-eighth of the country's agricultural land, damage 8,000 kilometers of roads, threaten the major port of Mongla, and require the resettlement of communities living in the entire region of Khulna, with the resettlement alone costing $13 billion.[32] Table 2.2 shows how climate change affects virtually every area of social, political, or economic activity in Bangladesh. Taken collectively, these factors combine to make Bangladesh exceptionally exposed to the impacts of climate change.

Table 2.2 Areas and sectors vulnerable to climate change in Bangladesh

Climate & Related Elements	Critical Vulnerable Areas	Most Impacted Sectors
Temperature Rise and Drought	North West	Agriculture (crops, livestock, fisheries), water, electricity supply, health
Sea Level Rise and Salinity Intrusion	Coastal Areas, Islands	Agriculture (crop, fisheries, livestock), water (water logging, drinking water), human settlements, electricity supply, health
Floods	Central Region, North East Region, Char Land	Agriculture (crops, fisheries, livestock), water (urban, industry), infrastructure, human settlement, health, energy
Cyclone and Storm Surge	Coastal and Marine Zone	Marine fishing, infrastructure, human settlement, life and property
Drainage Congestion	Coastal Area, South West, Urban	Water (navigation), agriculture (crops)

Source: Adapted from Government of Bangladesh (2005). *National Adaptation Program of Action* (NAPA). Dhaka, Bangladesh: MOEF.

3 Exposing the political economy of Bangladeshi adaptation

Because of this extraordinary susceptibility to climate change, the Government of Bangladesh, with the support of development partners and private actors, has invested billions of dollars over the past few decades to manage disaster-related risks. These investments have included flood management schemes, coastal polders, cyclone and flood shelters, and the raising of roads and highways above flood levels.[33] However, as a response to the country's increased vulnerability and the severity of cyclone *Sidr* in 2007, the government – for the first time – developed and implemented an integrated climate change strategy and action plan, based on its National Adaptation Program of Action (NAPA), in 2008. A local fund of $100 million was established exclusively for local climate change efforts.

Currently, the Ministry of Environment and Forests (MOEF) is tasked with monitoring and managing climate change affairs. The government has established an inter-ministerial committee on climate change, headed by the MOEF, comprised of representatives from relevant government ministries and departments as well as key NGOs and research institutions. The Department of Environment, under the MOEF, has also set up a Climate Change Cell to act as Secretariat for climate change-related work within the government. There is a National Environment Committee, chaired by the Prime Minister and with representation from Members of Parliament as well as government and civil society. To coordinate these many actors, the national climate change adaptation plan attempts to implement a "pro-poor climate change management" strategy, prioritizing adaptation and disaster risk reduction. It also aims to address national concerns with respect to climate change, including food security, social protection and health, comprehensive disaster management, infrastructure, research and knowledge management, mitigation and low carbon development, and capacity building and institutional strengthening.[34]

The centerpiece of these efforts is the country's NAPA. National Adaptation Programs of Action are documents registered with the UNFCCC detailing the urgent and immediate adaptation needs of least developed countries. They are intended to be peer-reviewed sources of information that provide an objective assessment of ranked climate change adaptation needs and vulnerabilities, as well as sector-specific costs and benefits.[35] Bangladesh's NAPA, in particular, calls for an integrated set of fifteen specific adaptive measures, with the following five ranked in order of importance:[36]

1. Coastal afforestation
2. Expansion of drinking water supply systems
3. Land water zoning and better water management
4. Improved emergency and disaster preparedness systems
5. Construction of flood shelters

The primary actor in charge of the NAPA process, the MOEF, notes that it generated this list of prioritized actions after screening more than 40 different options and consulting with more than 100 stakeholders.[37]

Though these five efforts look good on paper, in practice they succumb, in part or in whole, to the four political processes we describe in the book. To refresh your memory, the process of enclosure is when actors use climate change adaptation as a means to funnel public assets into private hands. Exclusion is when communities see adaptation projects limit their access to resources or become marginalized in decision-making processes. Encroachment is when adaptation projects damage the environment and intrude upon forests, farms, or indigenous lands, and entrenchment is when adaptation projects worsen social inequality. The remainder of this section elaborates on each in turn.

3.1 Enclosure

Adaptation projects intended to help coastal and predominantly poor populations have instead, at times, been plagued by land predation and land grabbing. *Khas* (public) and *char* (coastal island) lands are most at risk. To understand the process of enclosure more fully in Bangladesh, it is first important to provide a bit of background on the history of land tenure. Interestingly, the process of land grabbing in Bangladesh is both rural and urban, and it has been characterized as consisting of at least two categories, land seizure and land denial, and two processes, *ex situ* and *in situ* displacement.

For instance, many rural riverine and coastal sediment regions, constantly shifting islands called *char* lands, are "contested sites ripe for power plays that uproot small producers on their rich alluvial soils."[38] Many urban lands are captured by elites who use corrupted public servants, the military, and even gangs carrying bamboo sticks – thus the chapter title "Bamboo Thumpers" – to forcibly coerce small landowners to relinquish the titles to their property. When an area already owned or in possession of a group is taken over by others, it is known as land seizure. When a group is prevented from acquiring or accessing land to which it is entitled, it is known as land denial.[39] Lastly, *ex situ* displacement is a process whereby people are directly and forcibly removed from their land; *in situ* displacement is when struggles for or regulation of land indirectly leads to expulsion, such as through higher prices or changes in the law.[40]

Amazingly, in Bangladesh multiple actors are engaged in land seizure and denial. As one study documented:

> The most active agencies have been government departments and forces as well as private interest groups inclusive of commercial land dealers and speculators, political power holders, and civil and military officials in their personal capacities. In some parts of the country, land grabbing has been driven by socio-political factors related to sub-national political and ethnic conflicts.[41]

Thus, many actors have been quite effective at preventing many Bangladeshis from owning any type of land. One study estimated that roughly half of the rural population, and three-quarters of the country's total population, is landless or holds such a small amount of land that they cannot cultivate crops.[42] Another study projects that of all public land declared or set aside for its intended users, such as poor peasants or indigenous communities, only 11.5 percent has been effectively or legitimately disbursed.[43] As a joint report from the Human Development Research Centre and the Association for Land Reform and Development warned, "Bangladesh has a sizeable population of indigenous communities... [who] are losing their lands gradually because of lack of proper documentation, official records, and grabbing by others."[44] That same study also estimated that only five percent of fishing areas have been leased to poor fishers, with the rest (95 percent) going to "waterbody-grabbers."[45]

Trends, furthermore, are getting worse rather than better. Another report from the International Institute for Environment and Development revealed that "large numbers of Bangladeshis suffer landlessness, and despite land reform legislation and the abolition of intermediary landlords, land distribution has become more unequal in recent years, as a result of substantial problems of poor governance and corruption in the management of public land."[46] Climate change, perhaps understandably, will only exacerbate the situation, especially for *char* lands, which can disappear temporarily or permanently or reappear due to alterations in river flow, storm surge, or sea level rise.

Unfortunately, adaptation projects themselves can become conduits for enclosure and land grabbing. Perhaps the most direct link concerns the country's attempts to promote afforestation. In addition to being part of the measures covered by Bangladesh's NAPA, these actions also fall under a Coastal Development Strategy (CDS), the "lynchpin" of Bangladesh's Integrated Coastal Zone Management process, which is an attempt to "optimize the use of coastal land" to reduce the risks of climate change.[47] Though well intentioned, the CDS, which was implemented from 2006 to 2010, was criticized for increasing demand for land since it decreased the availability of "protected" property for agriculture. Moreover, property falling under the CDS was still subject to the usual government distribution of *char* lands, and "tyrants" and "mostly rich" people "illegally grabbed" the land and then excluded others from accessing it.[48] Further undermining the success of the CDS, issues and concerns related to climate change adaptation were not mainstreamed within respective sectoral planning institutions. As a result, the victims of climate change were underserved by their weak government. As Rashid writes:

> Core targets of Sixth Five Year Plan (2011–2015) of Bangladesh have been set in the context of Vision 2021. This plan acknowledges that climate-induced coastal submergence will be one of the major causes of population displacement. Therefore, effective steps must be explored and adopted in collaboration with the international community within the Sixth Plan period to mitigate problems. This plan regretfully mentions that the poor have very little access to government land like *char* lands, *khas* lands, water bodies etc. There are land laws and policies to allot such land to the poor and the landless, but in actual allocation, the interest of the poor is rarely preserved.[49]

Weak governance explains, in part, why after the CDS ended, coastal villagers reported no major increase in resilience or reduction in poverty.

Other interactions between climate change adaptation and land grabbing are less direct, but no less potent. Although these might not necessarily correspond to climate change adaptation measures, efforts intended to boost agricultural output or lower poverty have given rise to intensified land predation by elite members of Bangladeshi society. As the International Institute for Environment and Development explained:

> While the complex bureaucratic procedures for land allocation are already inaccessible to the landless, spontaneous land grabs occur before official surveys can take place. Land grabs of *char* land are frequently organized by powerful landlords (jotdars) under a neo-feudal system using dependent peasants (lathiyals), frequently people made landless by flooding, who may be formed into "cooperatives" registered under the patronage of the jotdars and given preferential treatment in subsequent land allocation by the landlord.[50]

In sum, elites use the process of climate change, and their knowledge about where adaptation projects will be implemented, to enclose upon land that they hope will be valuable. Researchers at the University of Dhaka in Bangladesh have shown that some well-connected stakeholders are able to use government divestiture of land for climate change projects as a mechanism to "gobble up" property and custodial rights.[51]

Despairingly, once elites have grabbed land, they tend to hold onto it with a tight fist, backed with violence and even financial brokers and members of the court. After land is claimed, land grabbing groups frequently hire their own small militia or police force to establish security and tenure over their land. These *lathials*, literally "bamboo stick-wielders," remain on the property for up to a few years to challenge any attempts at reclamation. As time passes, claims by the grabbers for formal property rights gain credibility, and ownership is enforced by the courts – meaning force and coercion are no longer needed.[52] Landholders will often rent out their newly acquired land to other poor farmers or sharecroppers who come to depend on it for their livelihood[53] – a perverse and perhaps ironic twist of fate. People, even the villagers themselves who might have been initially displaced, come to accept the land grab as legitimate. Similar systems of patronage – where bandits or criminals seize land and then use violence or force to keep it until it slowly becomes normalized as "theirs" – have also been documented in Bangladeshi forests.[54]

3.2 Exclusion

We see the process of exclusion also affecting multiple dimensions of climate change adaptation in Bangladesh, ranging from exclusionary forms of planning or implementation at the national scale to elite domination at the community scale.

At the national scale, the preparation of key climate documents and the collection of data on climate change have, in some instances, been elitist and exclusionary. For instance, the Bangladeshi NAPA process included economists and scientists as well as government officials, but not representatives from the most vulnerable groups, their professional associations, and civil society organizations.[55] Furthermore, though *char* water bodies are an important resource for fishers, national policies exclude *char* dwellers from land auctions, thus giving elites privileged access.[56] Lastly, authorities have frequently refused to prosecute or pursue criminal charges for the land grabbing activities mentioned above, dismissing claims from victims as frivolous and limiting access to legal recourse.[57]

On the implementation side, ethnic or religious minorities and women have been excluded from decision-making, even when communities try to implement components of the NAPA. To those who have lived in Bangladesh, this might come as no surprise, given that the country remains elitist in its local political and economic structures, and heavily patriarchal in it social structure.[58] Climate change actions in Bangladesh do not enter neutral territory. Instead, one study cautioned that "the archetypal situation in today's Bangladesh villages is elite repression and non-elite compliance;" that is, local elites come to dominate others and utilize resources to retain their positions of power.[59] In some parts of Bangladesh, such as *char* lands or along rivers and coasts, the system is almost feudal, where a "system of relations between landlords and peasants" exists, where "the local *talukdars* (independent proprietors) and *jotedars* (superior tenure holders), acting as power bosses, use their patron-tied dependents as *lathiyals* to organize violent land conflict."[60] The poor and powerless, in other words, all too often have little to no voice in village or community decisions of importance relating to adaptation.

A final type of exclusion is often both ethnic and sexist: social customs in Bangladesh dictate that members of the landholding class, mostly upper caste Hindu women, refrain from manual labor. That is, they "almost never work in the fields."[61] Instead, it is lower caste men that do the bulk of the physical work. This has grave implications for community-based adaptation schemes because it suggests that much of the physical effort of digging canals, erecting flood levees, and planting crops will be done by low-wage laborers with few political and social rights or, perversely, little access to the fruit of their labor when completed.

3.3 Encroachment

Because of Bangladesh's heightened vulnerability to climate change, the environmental impacts from its adaptation projects can become quite large, leading to encroachment upon other predefined land uses relating to coastal beaches, *char* land, forests, farms, and other public commons.

The most significant, capital-intensive intervention has focused on the erection of polders, designed to provide protection against storm surges, flooding, and tidal intrusion. A national Coastal Embankment Project has existed since the 1960s, and it constructed a staggering 5,100 kilometers of embankments and sea dikes to form 123 polders protecting 1.5 million hectares of land. When these polders began to need retrofitting and repair due to greater intensity and severity of storms, and drainage congestion problems, a Coastal Embankment Rehabilitation Project upgraded existing infrastructure and then extended it.[62] Most recently, in 2013 the World Bank launched the $400 million Coastal Embankment Improvement Project, intended to rehabilitate 600 kilometers of embankments in 17 polders in six coastal districts – Bagerhat, Khulna, Satkhira, Barguna, Patuakhali, and Pirojpur – and to protect 760,000 people living within the polder boundaries.

Today, the Bangladesh Water Development Board manages an intricate network of 9,943 kilometers of embankments, 5,111 kilometers of drainage canals, and about 14,000 flood control regulating structures with a total cost of about $2.7 billion.[63] Perhaps obviously, these types of efforts encroach upon private uses of land. The Coastal Embankment Improvement Project, merely an upgrading of embankments, is expected to displace more than 6,200 households physically or economically, destroy more than 2,000 businesses, damage 1,315 agricultural plots, and force the closure of 184 educational and religious buildings.[64]

A secondary problem is that these infrastructural measures provide unequal protection. As Feldman and Geisler write:

> For several decades the Government of Bangladesh has sought to stabilize such lands by constructing polders and flood control structures, projects largely supported by the donor community. But, under these schemes as well, the benefits often accrue to the more wealthy and secure members of the community, including through the corruption that accompanies their very construction.[65]

Put another way, the construction or rehabilitation of climate-proofing infrastructure has encroached frequently upon the land held by poor, marginalized coastal inhabitants and protected wealthier stakeholders further along the coast. Its construction also siphons off resources, through corruption, to workers and laborers associated with local elites.

Other encroachment is related to the conversion of *char* land and forests. For example, according to national climate policy, as soon as coastal *char* land is located, it is handed over to the Forestry Department, which serves as its custodian for at least 20 years. Rather than circulate this land back to people, mangrove forests are planted as part of a coastal protection measure. Only after accretion and consolidation of the silt intensifies beyond a certain point, often taking years or even a decade, is the area behind the mangrove belt converted into land for human settlement.[66] Similarly, a national Coastal Afforestation Scheme ran for more than three decades and placed roughly 35,000 hectares of forest plantations under national protection as part of a "Coastal Green Belt," understandably excluding local villagers or communities from cultivating the land so as not to compromise its ecological integrity. In tandem, a National Land Use Policy intends to bring one-quarter of all land across the country under forest cover and advocates conservation of existing forested parks, including the Sundarbans.[67]

A final type of encroachment relates to environmental degradation from industrial farming, promoted by the government as a measure to build economic development and resilience among coastal zones. The government launched the *Chingri Mahal*, or shrimp zone rules, in 1992 in an effort to lure foreign investment and generate "high economic returns."[68] These rules were followed more than a decade later by the Coastal Zone Policy in 2005 and a Coastal Development Strategy in 2006.[69] These latter efforts sought "to create conditions, in which the reduction of poverty, development of sustainable livelihoods and the integration of the coastal zone into national processes can take place."[70] More specifically, the top two objectives of these two policies were to "promote economic growth" and to enhance "basic needs and opportunities for livelihoods." All good stuff. Unfortunately, the implementation of these plans has rested upon the establishment of large-scale industrial enterprises such as shrimp farming. Such activities have transformed previously public agricultural or *khas* lands into sites of industrial output owned by national and, at times, transnational corporations. Part of this encroachment is about integrating "smallholder agriculture" into global commodity chains;[71] part is about boosting export earnings and attracting foreign direct investment.[72]

Regardless of their intended purposes, the result of the shrimp zone and coastal development policies has been priority allotment of state lands into private hands. As one study explains, "while the erstwhile priority given to poor peasants continued to be nominally kept in the books, wealthy interest groups became entitled to claim priority in land allotment by invoking the official rules for promotion of shrimp farming."[73] Consider the situation at Polder 23, a 5,800-hectare polder in Khulna's Paikgacha *Upazilla* (sub-district), where 84 percent of residents are landless as a result of the expansion of commercial shrimp farms. As Yale University researchers Paprocki and Cons write:

> Prior to the incursion of shrimp into the polder, residents claim that it was possible even for sharecroppers and day laborers to achieve household self-sufficiency by combining wage labor with farming on the polder's khas (common) land. Now, the majority of land within the polder, including khas land, has been overrun with shrimp. As a result, residents report not just a decline in the availability of nutritious foods, but a shortage of labor opportunities, an inability to pay the fees necessary for sending children to school and a marked increase in indebtedness both to local moneylenders and to microcredit organizations.[74]

When asked how they felt about the effect industrial shrimp aquaculture had on their community, villagers used words such as *dhongsho, bilupto, shesh* (destroyed, extinct, finished).[75]

3.4 Entrenchment

A final political economy dimension to adaptation in Bangladesh is entrenchment, the aggravation of disempowerment or increased income inequality across communities. Like our other factors, this one also has complex linkages and effects, but we document how coping strategies for climate change are entangled with existing class and ethnic hierarchies that not only disseminate the benefits of adaptation unevenly, but trap many of the poor, powerless, and displaced into a dangerous patrimonial system of insecurity and violence.

At the start, community coping strategies can worsen gender inequality. One case study of displacement following rapid riverbank erosion in *Char Nalgonda* noted that most families had to relocate five times and some families shifted their homestead 25 times since originally settling in the region.[76] This process inequitably distributed work to women. Women in the household were required to construct and raise the household mud platform (*biti*) and make a new oven (*chula*) and establish a new garden, and, if desired, build new basic infrastructure such as tubewells and latrines.[77]

Beyond gender, issues of class, ethnicity, and village hierarchy come into play. After a severe flood occurs, the typical response from villagers is to seek social and economic support from friends and local institutions before they approach NGOs, aid donors, or the government. Given the paucity of state resources in Bangladesh, this makes sense, but it also means that victims become more susceptible to local political economy processes, including profiteering. For instance, one study of refugees from flood and riverbank erosion in Bangladesh found that many villagers had to sell their cattle and household utensils to meet simple needs such as water and food.[78] Another examination of land expropriation in Bangladesh noted that many *char* residents had to abandon homestead land, housing materials, crops, cattle, and even trees when confronted with a disaster, selling them at cut-rate prices to local marketers.[79] A report from the University of Dhaka comments that after a national or local disaster, "local elites and influential groups take advantage of the situation because local marginalized farmers in many cases are forced into distress selling."[80] The implication here is that victims can face entrenchment not necessarily by private corporations, but by their own neighbors.

Displaced households lacking savings or property are forced into an even worse situation: they must go into debt, migrate, or starve. One study of coastal hazards and community coping methods in *Hatia*, an island in the Noakhali district, found that those without property, savings, poultry, or livestock – a strong majority – had to take a loan from relatives, or, a more common practice, moneylenders.[81] According to Parvin, Takahashi, and Shaw, those who could not secure loans tended to leave the community in search of jobs in urban areas, usually Chittagong or Dhaka.[82] If a job could not be found, those without access to alternatives were usually "forced to resort to starvation, sometimes only one meal a day, or nothing at all."[83] Another assessment of coping strategies in the Brahmaputra-Jamuna floodplain noted, again, that the displaced are dependent upon their kin and local groups for assistance, and in the absence of those, they need to rely on their friends and relatives, or without those, relocate and starve. As the authors concluded, "displaced households having no land of their own to resettle, or who cannot expect any material assistance from their equally poor relatives, have very limited choices" and "adjustment to displacement in the active floodplains of Bangladesh has been conditioned historically by social, cultural and political factors."[84]

Sordidly, an elaborate patronage system has arisen to exploit this process of displacement. Many of the displaced persons without any alternative seek permission and protection to remain in *char* areas as the dependent of a village faction leader or a locally powerful chief, a *matabba*, or a locally powerful boss (called a *talukdar* during colonial times). These local elites provide necessary support for resettlement – in some cases food and water, in other cases temporary land – because the landless displaced represent "a cheap labor pool to work on their land."[85] Some of the displaced are also useful as loyal *lathials*, or bamboo-stick-wielding hoodlums promoting the process of enclosure discussed previously. In many cases, the roles of law enforcement agencies are coopted by local elites and their affiliated political parties.

Such a system, despite its viciousness, seems to be particularly widespread in Bangladesh's most vulnerable areas. Feldman and Geisler write that:

> Although the incidence of destitute displacees is quite high in the region, relief and rehabilitation programs are seldom undertaken by national government or nongovernment agencies or by local government offices...This assistance vacuum and the limited options for displacees brought about an insidious dependence on local patronage systems and set in motion...multi-staged displacement...Impoverished displacees become a cheap labor pool of sharecroppers and agricultural workers for local talukdars (bosses) and serve, as well, as violence enforcers of land grabbing for one or another local faction. That is, they are transformed into "readily available lathiyals (clubmen, armed retainers) to organize violent fights when necessary to gain control over newly emergent char land." This perverse process, turning displacees into displacers, is not an isolated incident.[86]

According to one survey, more than 50 percent of *char* villagers are displaced and landless "dependents" embroiled in seedy patronage networks.[87] Another study of forced migration in Bangladesh noted that in situations where refugees do not have family or savings, landlords almost always demand agricultural labor, household service, and political support in exchange for a house plot.[88]

Life as one of these landless, powerless, displaced people is far from enjoyable, to put it mildly. For instance, Abul Kalam, who used to be a professional fisher, is now a rickshaw puller living with his family of eight in a temporary thatched shack next to the canal of a fishery *ghat* (landing center for boats). He migrated to Chittagong city in 2008 after he lost three acres of land due to the erosion of the Meghna. As he says:

> I shifted house three times due to erosion. My family members lived on other land after losing assets. Erosion changes everything; our home, livelihood and the society as well. River erosion is the curse for us.[89]

He has four sons and two daughters. Before their displacement, his eldest son was a day laborer in a village, but now he subsists by skinning fish. Abul's second and third sons, and eldest daughter, dropped out of school and are currently unemployed.

This isn't the worst that can happen. Adnan researched land struggles in the *Noakhali* coastal belt, where landless and displaced peasant squatting on *char* lands is common. As he explains, these people:

> [f]aced recurrent violence from rival power-holders in the form of arson, harassment, assault, rape and killings, which were usually linked to threats of eviction from the lands under their occupation. Furthermore, these migrant households were subject to the absolute social power of the banadasyu leaders within their respective domains, who "promulgated local laws" and used violence and intimidation to exploit and repress them. They had little option but to suffer in silence since they needed the protection of their own banadasyu leader against attacks from rival power-holders seeking to grab their tenuous landholdings.[90]

In the face of climate change, land tenure and human security for the displaced usually depends upon pledging loyalty to one criminal boss or another. The system of patronage is inescapable. Further, all of the displaced have political value to the *matabba* and *talukdar* as captive voters in local and state elections, ensuring that the mechanisms of entrenchment only serve to perpetuate themselves.

4 Uncomfortable conclusions and implications

We offer four broader conclusions for climate change adaptation policy and practice, both in Bangladesh and the world at large.

First, within Bangladesh, we see the presence of not just one element of our political economy framework but all four – enclosure, exclusion, encroachment, and entrenchment – acting at multiple scales across various stakeholders. As Table 2.3 summarizes, the political economy of

Table 2.3 The multi-scalar nature of the political economy of adaptation in Bangladesh

Scale	**Political Economy Predominantly Affects**	**Explanation**
Macro: National export zones, industrial clusters, and protected areas	*Khas* land, farms, coastal villages, forests	People become displaced or lose livelihood opportunities in the face of climate-infrastructure, forest conservation areas, waterways, and aquaculture sites
Meso: Cities and regions	Public commons, *char* areas, hilly agro-ecological regions	Powerful landlords (*jotdars*) utilize a neo-feudal system of dependent peasants (*lathiyals*) to consolidate power and concentrate wealth
Micro: Neighborhoods and households	Family assets, *banadasyu* patronage networks	Moneylenders and community leaders take advantage of disaster victims by forcing them to sell assets at low prices or recruiting them into a patronage system managed by the *matabbar* and *talukdar*

Source: Authors.

adaptation transcends micro, meso, and macro scales with distinct underlying causal factors. At the macro scale, national policies have reoriented efforts toward boosting resilience and enhancing exports and economic development, practices that protect some – notably wealthy landowners and shrimp farm industrialists – but exclude others – notably the landless and displaced peasants. At the meso scale, at the level of cities and regions, we see thugs and bandits roaming the countryside to steal land or appropriate resources, with an eye for which climate change adaptation projects protect and make land more valuable and worth grabbing. At the micro scale, within neighborhoods and households, we see how family members cope with disaster, often by delegating some of the hardest work to women or ethnic minorities, or by taking advantage of distress selling by households that lack savings or property.

Second, many of the political economy elements we identify in this chapter are not entirely the result of local forces or factors. Instead, we see actions in Bangladesh connected to global processes such as the funneling of international donor money (e.g., hundreds of millions of dollars from the World Bank) into national infrastructure projects such as polders, embankments, and canals. We witness the desire to attract foreign direct investment driving the expansion of industrial shrimp farms. Most of all, we see a global demand for cheap labor and commodities placing pressure on national systems of Bangladeshi governance, and local systems of village hierarchy and patronage, that place profiteering and accumulation of wealth before the welfare of individuals, particularly the landless and displaced.

Third, because of its political economy, some of the adaptation interventions undertaken in Bangladesh will require further adaptation. The more than $3 billion placed into the construction or rehabilitation of polders will require costly maintenance and wide-ranging resettlement and relocation plans. Moreover, if sea levels rise beyond the height of these protective measures, they could even trap flood or storm surge water behind them, backfiring and worsening the impacts of climate change. The industrial shrimp farming and export-oriented economic development ostensibly intended to alleviate poverty have damaged river embankments and facilitated the construction of sluices that allow saline water to enter freshwater areas, leading to the intrusion of brackish water that makes cultivation of other crops difficult.[91] Community coping strategies predicated on property and loans can become a stark liability when those assets run out, placing family members in a predatory system where workers pledge economic and political loyalty to clan bosses in exchange for shelter, food, and security.

Fourth, the presence and pervasiveness of some of the political economy elements we identify with adaptation – especially the classism, violence, and feudal nature of patronage – should serve as a wakeup call for Bangladeshi planners that they can no longer ignore the broader social and political environment in which they operate. For, while it is true that climate change can be described as a major causal factor creating "descents into poverty" where households succumb to flooding and ill-health,[92] our research suggests that community responses, and adaptive measures, can also serve as a lever that forces Bangladeshis to descend and remain trapped in poverty and human insecurity.

3
Degraded Seascapes: The Political Economy of the Dutch Delta Works

1 Introduction

To some, the topic of storm surge – rising water commonly associated with low-pressure weather systems such as tropical cyclones – might sound mundane, but it could pose a colossal threat to humanity by the end of this century. For instance, one team of researchers found that, due to climate change, average storm surge damages were likely rise from $10 billion to $40 billion in 2014 to possibly $210 *trillion* ($2100,000 billion) by 2100, representing as much as 9.3 % of global GDP and affecting up to 600 million people.[1] Low elevation coastal zones around the world cover only about two percent of landmass but contain more than ten percent of the human population.[2] A team of Chinese researchers projected that no less than 57 countries will face significant storm surge impacts by the end of this century.[3] Those researchers estimated that Australia and the United States will exhibit the greatest inundated areas; Bangladesh, India, China the most affected populations; and the United States, China, Japan, and Mexico the greatest economic losses.[4] An article in *Nature* warns that "climate change and sea-level rise present major challenges to each of the world's delta regions, which together harbor about 70% of the world's population and economic resources."[5] Indeed, by combining future global sea level rise with tide gauge water levels, another research team expects that today's "once in a century" storm surges might become "once in a decade" storms in the future.[6] Still another study predicts that insured losses due to storm surges will rise by about 20 percent across all parts of the United States between 2010 and 2030.[7]

Because so much is at stake, and because storm surges pose a complex, persistent, and magnifying threat,[8] elaborate plans have arisen for coastal protection infrastructure, ranging from dikes and polders (see our chapter on Bangladesh) to river barrier systems (such as the one controlling the Thames near London, and outside Venice), and tetrapod seawalls (such as those protecting parts of the Maldives). In the United Kingdom alone, engineers expect that £22–£75 billion of new infrastructure will be needed to manage storm surge flooding risks.[9]

Perhaps no country is more at biophysical risk – defined as probability and magnitude[10] – to storm surge, however, than the Netherlands. More than half of the country's inhabitants – some 8.5 million people – live below sea level, and two-thirds of the country's landmass is vulnerable to perpetual flooding. As one study put it, in the Netherlands, "the battle against the North Sea is widely recognized as a safety issue of continuing national importance."[11] Therefore, engineers have erected an elaborate system of storm surge barriers and dikes intended to protect against a 1 in 10,000 years storm surge event for the provinces of Holland, and 1 in 4,000 years storm surge event for the provinces of Zeeland, Friesland, and Groningen.[12]

This chapter begins by offering a brief history of Dutch flood protection before detailing the genesis of the Dutch Delta Works, including more details about its centerpiece, the Eastern Scheldt Storm Surge Barrier. The chapter proceeds to illustrate how the Delta Works suffered from, or perpetuated, our four political economy processes of enclosure, exclusion, encroachment, and entrenchment before offering four conclusions.

Now, to be fair, we wanted to mention that to a significant degree, the Dutch Delta Works was an artifact of its times. It was conceived and then initiated many decades ago, before the birth of both of the authors. Dutch planners were facing a serious, aggravated threat of future storm surges and flooding and were recovering from a major national crisis (the 1953 flood). Their intentions were venerable – protect people first – and they were working prior to the development of many modern principles of governance, stakeholder involvement, and environmental sustainability. Moreover, the principal agent of the Delta Works, the *Rijkswaterstaat*, has seen itself institutionally reformed in significant ways since the completion of the projects, becoming restructured, smaller in size, and much more inclusive and transparent.[13] It is in one sense unfair to evaluate past decision-making processes through the lens of today's best practices. We still do so in this chapter because, while we appreciate the goals and strategies enshrined within the Dutch Delta Works, our point is that such long-lived infrastructure begins to carry out a life of its own, a life that must be managed and accommodated decades after the project is completed. In short, the Delta Works carried its 1950s and 1960s style values well into the future.

2 Dutch flood protection: A prelude to the Delta Works

In the Netherlands, where at least one-third of the country's land lies below sea level and about two-thirds of gross national product is earned in flood-prone areas,[14] defensive measures against storm surge, sea level rise, and flooding have been both necessary and long-lived.[15] Much of the modern-day nation is habitable by humans only via the "incessant renewal and maintenance of a complex system of waterworks that keep the sea and rivers at bay."[16] It is a system that has been continuously preserved for more than one thousand years. Because of this complex protective system, Bijker comments that "living in low-lying areas protected by dikes, often below sea-level, is an accepted fact of life in the Netherlands."[17] As the locals often joke, "God created the world, and the Dutch created the Netherlands."[18]

Indeed, the first forms of democracy and governance during the Middle Ages in the Netherlands were reputedly related to the management of dikes and sluices. Nature, especially the sea, was seen as the "hereditary enemy," one which took lives and property maliciously "whenever and wherever it could."[19] Until roughly the ninth century, land was continually lost every year due to natural accretion and erosion.[20] From the thirteenth century onwards, however, farmers started resisting and offsetting land loss with small dikes intended to keep areas that dried at the low waters level permanently dry.[21] This was followed by the invention and implementation of polders, low-lying areas prevented from flooding by a heightened dike.[22] After the sixteenth century, Dutch workers began consistently making polders by digging drainage ditches and canals, building dikes alongside rivers and vulnerable coastal areas, damming tidal bays, and reclaiming land through mechanical pumping.[23] The process worked, and not only stopped the Dutch from losing land, but slowly allowed them to regain thousands of hectares each century, culminating with the most intricate system of levees, dikes, and polders the world had ever seen in place by the turn of the twentieth century.

However, the twentieth century also saw significant threats emerge. Dikes, in particular, started to fail in the early and mid-1900s for at least three interconnected reasons. One problem was that comparatively poor construction materials, notably old sand and clay, finally started to give way to the elements. Until the 1930s and 1940s, little was actually known about what factors really influenced the strength, or failure rates, of a dike. Because there was no alternative, dikes were placed on unstable subsoil or peat, leaving them vulnerable to the effects of erosion. Due to economic considerations related to preserving as much land as possible, landward sides were made too steep, leading to premature cracks or erosion. Seaward faces were generally covered with stones or basalt, landside surfaces with grass, causing liquefaction of loosely packed sand and leading to an eventual sliding and collapse of walls, a process known as overtopping. Between 1900 and 1940, for instance, overtopping occurred at some 140 dikes.[24]

Another problem was that despite their social importance, the dikes were poorly cared for. As one study notes, before 1945, "hardly any merging of polder boards took place, and there was no overall approach to dike maintenance."[25] Before and after World War II, Dutch planners understandably prioritized other policy issues such as national security or education. Given that they believed that the risk of a future major storm was low, supervision of dikes remained largely "on paper" only, and rehabilitation work under the auspices of the *Rijkswaterstaat* (the national public works organization) or the *Provinciale Waterstaat* (its counterpart at the provincial level) was exceptionally rare.

A final problem was that defense work undertaken for World War II weakened the state of poorly made, poorly maintained dikes even further. Military bunkers, and in some instances entire defense complexes, were built into the dikes themselves, as part of a national security system attempting to expel an attack from the sea. This included machine gun units, manholes, and piping that literally cut through dikes, allowing water to enter. After the war, many of these weak spots were never sufficiently repaired, and their degradation was worsened by the work of rabbits building warrens and moles building nests.

Thus, though the nation didn't necessarily know it, the Netherlands was extremely vulnerable to the severe storm and resulting flood that occurred on February 1, 1953. A combination of high tides, strong waves, and severe winds resulted in the failure of a massive number of dikes – at least one hundred collapsed and at least 150 were breached in some manner. Roughly 130,000 hectares of low-lying farmland was flooded with saltwater, and 1,800 people were killed – almost the same number that died as a result of Hurricane Katrina in the United States (see Chapter 4). In addition, property damage was extensive, with more than 46,000 farms and buildings destroyed or damaged, and approximately 200,000 farm animals drowned.[26] Moreover, 100,000 people were evacuated, and 750,000 people were eventually negatively impacted in some way by the storm.[27] Figure 3.1 illustrates the storm's damage on a map.
Surprisingly, the flood could have been much worse if not for some quick thinking and a bit of providence. The dike located at the Rhine River

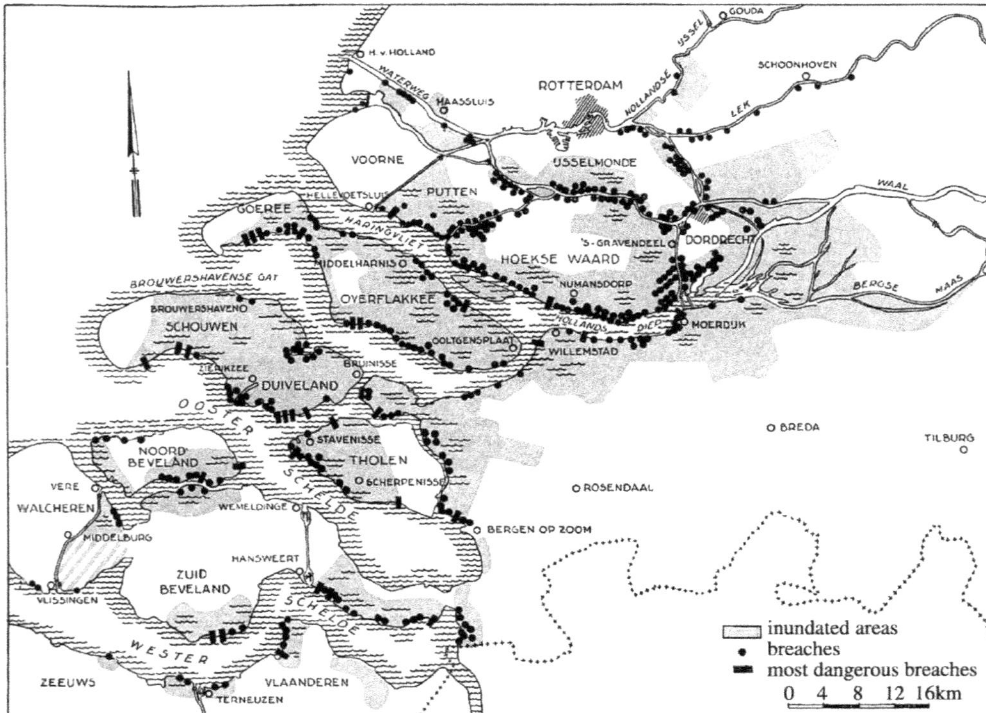

Figure 3.1 Dike breaches and inundated areas of southwest Netherlands, February 1, 1953

Source: Modified from Gerritsen, H. (2005). What happened in 1953? The big flood in the Netherlands in retrospect. *Philosophical Transactions of the Royal Society A*, 363(1831), 1271–1291.

at Ouderkerk was crucial to protecting central Holland and the urban areas of Rotterdam, Delft, The Hague, Leiden, Gouda, and Amsterdam. Amazingly, that dike was very close to being breached, but the Mayor of Vogelaar resolutely commandeered a nearby river freight ship and ordered its captain to maneuver it into the dike, breach it, and sink it – as Figure 3.2 shows. Citizens and solders then rushed to fill any remaining gaps with sacks of sand, concrete waste, and, in a few instances, household furniture. As one historical account surmises, "before nightfall the breach was closed and central Holland was saved. During that day, the old inner dikes held out, which was considered... sheer undeserved luck."[28]

To be sure, the 1953 storm had a lasting impact on both the national psyche and future coastal protection plans. One day after the storm, while the *Rijkswaterstaat* was still conducting rescue operations, it also began damage repair in concert with construction contractors and water boards. Within the month, the *Rijkswaterstaat* announced a plan of closing off dangerous estuaries so that the likes of that storm could never be repeated. On February 18, 1953, a mere seventeen days after the storm struck, the Minister of Transport and Public Works created the "Delta Committee" to provide recommendations on how flood-prone areas could be protected, and how the sea defenses could be strengthened along the coast of the North Sea. The Delta Committee published

Figure 3.2 The ship that saved the central parts of Holland from the 1953 flood

Source: Modified from Gerritsen, H. (2005). What happened in 1953? The big flood in the Netherlands in retrospect. *Philosophical Transactions of the Royal Society A*, 363(1831), 1271–1291.

five reports from May 1953 to October 1955, all of them supporting more elaborate, intensive physical protection from improved seawalls or storm surge barriers. These efforts culminated in the "Delta Project" or "Delta Works" (*Deltawerken*), a grand scheme passed by the Dutch Parliament on November 6, 1957. The main goals of the Delta Works were to completely close the Haringvliet, Brouwershavense Gat, and Eastern Scheldt estuaries, and to strengthen seawalls along the Western Scheldt, Rotterdam Waterway and other sections of the coastline.[29] The Eastern Scheldt Storm Surge Barrier, or *Oosterscheldekering*, was the centerpiece of this scheme.

3 History of the Dutch Delta Works

The massive Dutch Delta Works package, shown in Table 3.1, consisted of 14 projects spanning some 40 years at a projected cost of $10.26 billion.[30] The largest of these was the Eastern Scheldt Storm Surge Barrier, situated between the islands of Schouwen-Duiveland and Noord-Beveland, which protects a southwestern corner of the Netherlands known as Zeeland, a name appropriate since it means "sea-land." Zeeland is also sometimes referred to as the "Delta Region" since it is home to three major rivers flowing into the North Sea: the Rhine, the Meuse, and the Scheldt.[31]

Overall, the Delta Works were supported on the grounds that they would promote Dutch industrial innovation in construction and engineering, facilitate horticulture, provide a modest gain in land, and create freshwater basins for the drinking water supply.[32] The Eastern Scheldt Storm Surge Barrier is the largest part of these Delta Works, spanning nine kilometers and costing $4.7 billion to construct[33] (and about $25 million a year to maintain and operate).[34] The storm surge barrier's primary goal is flood protection and safety; it is designed to reduce the frequency risk of flooding to 1 in 4,000 years.[35] When it was completed in 1986, Queen Beatrix attended the opening herself, and christened it by saying that "The flood barrier is closed. The Delta Works are completed. Zeeland is safe."[36] Since then, it has been used (closed) dozens of times. Figure 3.3 shows the Eastern Scheldt Storm Surge Barrier in closed configuration during a storm.

The Delta Works formally began in 1957, and plans for many of its largest components, such as dikes and what would become the Eastern Scheldt Storm Surge Barrier, were almost continually discussed over the next two decades. For the Eastern Scheldt in particular, a draft proposal to begin construction was tentatively approved by Parliament in 1974, but

Table 3.1 Project components of the Dutch Delta Works, 1958–1997

No.	Date Completed	Name	Type	Purpose
1	1958	Hollandse Ijssel	Drop Barrier	Flood protection from the sea for the Rhine river
2	1960	Zandkreek	Tidal-Control Dam	Reduces tidal and current velocity for Veerse barrier, provides navigation to Lake Veerse
3	1961	Veerse	Fixed Barrier	Storm surge flood protection
4	1965	Grevelingen	Tidal-Control Dam, Lock, and Siphon Sluice	Reduces tidal velocity for Brouwers Barrier, facilitates navigation to Lake Grevelingen
5	1970	Volkerak	Waterway Dam and Navigation Lock	Storm surge flood protection
6	1971	Haringvliet	Sluice-Gate Barrier and Navigation Lock	Storm surge flood protection
7	1972	Brouwers	Fixed Barrier	Storm surge flood protection
8	1983	Marquisate Quay	Waterway Closure Dam	Part of the Scheldt-Rhine waterway link, reduces tidal velocity and controls saltwater intrusion
9	1986	Eastern Scheldt	Lift-Gate Barrier	Storm surge flood protection
10	1987	Philips Waterway Dam and Krammer Navigation Locks	Dredged Sand Embankment Closure	Part of Scheldt-Rhine compartmentalization closure
11	1987	Oester Waterway	Closure Dam and Navigation Lock	Facilitates shipping access to the Eastern Scheldt and controls saltwater intrusion
12	1987	Bathse Spuisluis	Lock	Reduces current velocity of the Volkerak, Markiezaatsmeer, and Oosterschelde Rivers
13	1997	Maeslantkering	Flood Barrier	Flood protection for Rotterdam Harbor
14	1997	Hartelkering	Storm Surge Barrier	Storm surge flood protection

Source: Modified from Watson, I. & Finkl Jr., C.W. (1990). State of the art in storm-surge protection: The Netherlands Delta project. *Journal of Coastal Research*, 6(3), 739–764. See also Delta Works Online Foundation/Deltawerken (2004). *The Delta Works*. Available at: http://www.deltawerken.com/Deltaworks/23.html.

Figure 3.3 The Eastern Scheldt Storm Surge Barrier

Source: Delta Works Online Foundation/Deltawerken (2004). *The Oosterschelde storm surge barrier*. Available at: http://www.deltawerken.com/The-Oosterschelde-storm-surge-barrier/324.html.

stalled due to concerns over the optimality of its design; that stall took two years to resolve. A modified plan was approved in June 1976, when the *Rijkswaterstaat* set up its first ever project organization consisting of three construction directorates and six large Dutch civil engineering contractors, a group colloquially known as the "Delta Directorate."[37] In late 1976, feasibility studies commenced at the same time that construction began in the western mouth of the estuary.

Construction was, for many reasons, complicated, time-consuming, and expensive. As Neinhuis and Smaal intone, "It soon became clear that the project was of unprecedented complexity."[38] At least five major interconnected reasons seem to account for its complexity and cost.

First, the project involved far more than the storm surge barrier itself. Complete protection of the estuary required a series of nine auxiliary compartment dams, topographically separated from the main seawall, along with a number of primary embankment dams, cofferdams, and secondary gravel and sand dams, collectively forming an "indissoluble entity" with the barrier.[39] Some of these dams, such as the Oesterdam shown in Figure 3.4, were large – 11 kilometers long – and capital intensive themselves.

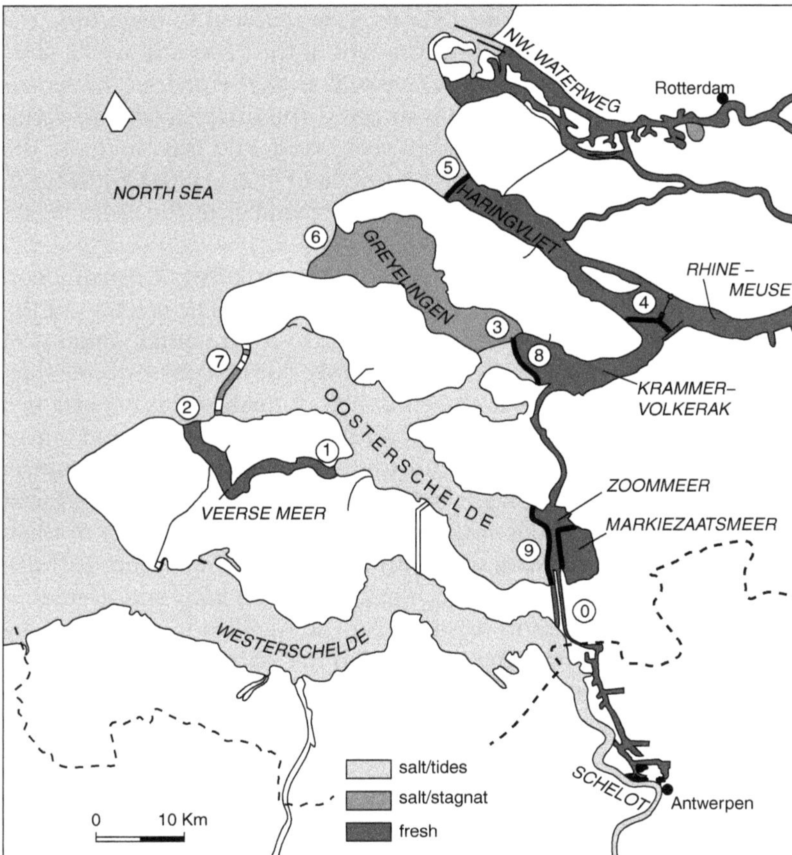

Figure 3.4 Map of the southwestern Netherlands after the completion of the Delta Works

Note: 0 = Kreekrakdam, 1867; 1 = Zandkreekdam, 1960; 2 = Veersegatdam, 1961; 3 = Grevelingendam, 1964; 4 = Volkerakdam, 1969; 5 = Haringvlietdam, 1970; 6 = Brouwersdam, 1971; 7 = Oosterschelde storm surge barrier, 1986; 8 = Philipsdam, 1987; 9 = Oesterdam, 1986.

Source: Modified from Nienhuis, P.H. & Smaal, A.C. (1994). The Oosterschelde estuary, a case-study of a changing ecosystem: An introduction. *Hydrobiologia*, 282/283, 1–14.

Second, the project required significant innovations in science and engineering – in physics, biology, ecology, materials, and modeling, to name a few. As civil engineers d'Angremond and Kooman note, "from the beginning of the Delta project, it was clear that the available knowledge and the traditional working methods for closing estuaries or dike gaps was insufficient" meaning "a program of basic research was therefore initiated."[40] Predicting future metrological conditions necessitated advances in computing power, information processing, and modeling. Complex calculations were required to determine the optimal form, composition, and strength of the dams and storm surge barriers as well as their impact on coastal areas.[41] Hundreds of experiments and simulations were run to assess the integrity and performance of building materials, natural rock, sand, and clay. In 1970, the *Rijkswaterstaat* created an ecology department in Zealand (Middelburg), where all state-of-the-art ecological impact studies of the Delta Works could be conducted.

Third, drawing from this new knowledge, a number of construction methods and vessels had to be specially designed. The erection of the main seawall, nine kilometers long, required a substantial amount of dredging and the creation of artificial islands. The first phase of construction, for example, required three months of dredging to convert two natural sand banks into permanent artificial islands, the second one of these serving as a convenient construction complex.[42] To accomplish the dredging, Dutch engineers invented the largest dustpan dredger in the world. Moreover, four ships had to be custom built, each of them named after different types of shellfish. The *Mytilus* was modified with various ground-working tools such as compaction guns and needles to make the seabed and sand more stable. The *Cardium* was rigged to lay a special carpet and mooring platform on the seabed to rest pylons on. The *Ostrea* was the first ship in the world capable of lifting 18,000-ton concrete pillars from a dry dock and placing them on the seabed. The *Macoma,* with specially designed sonar, enabled the *Ostrea* to accurately place its pylons.[43]

Fourth, due to a compromise to appease the concerns of fishers and environmentalists, the storm surge barrier had to possess the ability to open and close – it had to be a "semi-open dam" capable of fully closing when the magnitude of flood risk was high.[44] A series of 65 prefabricated concrete pipers (pylons) also had to be specially created to support 62 steel gates, which could be dropped, similar to the portcullises of castles, to close off the estuary during severe storms.[45] Perhaps amazingly, these elements only formed part of the skeleton of the Eastern Scheldt Storm

Surge Barrier – no less than 22 distinct foundations and structures made up the whole, including prefabricated support piers, armor stones, ballasted bases, and block mattresses.[46]

Fifth, despite all of the auxiliary infrastructure, basic research, engineering advances, and carefully laid plans, when construction began it became apparent that the seabed was not as stable as initially thought. Instead of being hard and solid, it featured soft, compressible clays and peats that needed excavating. Portions of the seabed were replaced with sand, and then covered with a layer of protective gravel. As one engineering team lamented, it was akin to "building on quicksand."[47] Large-scale "vibratory compaction equipment" and "compaction vessels" had to be deployed to de-muck and stabilize the seafloor before pylons and anchor piles for the seawall could be pounded in.[48] To protect the parts of the storm surge barrier exposed to seawater, unique polypropylene and concrete-block mats, asphalt slabs and graded-filter mattresses had to be invented and installed.[49]

For these reasons, both the Delta Works and the Eastern Scheldt Storm Surge Barrier suffered from delays and cost overruns. The Eastern Scheldt Barrier was finished some ten months behind schedule with a budget overrun of the storm surge barrier (alone) at 30 percent; the Delta Works as a whole were ten years behind schedule and had cost overruns reported at 15 to 30 percent.[50] That said, some actors saw it as a modern engineering marvel and worth the cost. The *Rijkswaterstaat* publicly declared that "the cost benefit analysis" of the project was "positive," meaning it was less expensive than the storm damage that would have occurred in its absence.[51] Within the Netherlands, the Delta Works became "celebrated as an expression of Dutch national vitality and Dutch civil engineering prowess in particular."[52] International reactions were equally buoyant. The American Society of Civil Engineers called it "one of the modern Seven Wonders of the World."[53] Another team of researchers proclaimed that the Barrier represented the "State of the Art in storm-surge protection."[54]

The perceived success of the Delta Works also meant that it would become exported as a model for future storm surge barriers around the world. The Eastern Scheldt Storm Surge Barrier grew into an exemplar – a first-of-a-kind engineering marvel to be replicated by other countries attempting to climate-proof their infrastructure.[55] Such strategies entail a paradigm of using "hard infrastructure" to reduce the risks of climate change "to a quantified level, accepted by the society or economy."[56] Assessing the Delta Works project, d'Angremond and Kooman concluded that its impact on the entire field of engineering was profound:

> The Eastern Scheldt storm surge barrier project has changed hydraulic engineering from a traditional activity to a scientifically based discipline. Despite the many large technical uncertainties, working conditions never met before, and demands on tolerances offshore which normally are only met on-shore in well-defined circumstances, the time target has been exceeded by less than one year ... Compared with other unique megaprojects, this is quite a success.[57]

As Table 3.2 indicates, the knowledge from the Delta Works has been used in at least five large-scale storm surge barriers totaling almost $11 billion of collective investment in the Netherlands as well as in Germany, Italy, the United Kingdom and the United States.

Moreover, after Hurricane Katrina hit the United States in 2005, the US Army Corps of Engineers visited the Delta Works and then awarded a $150 million contract to a group of Dutch engineering and construction companies for the management of levees and floodwalls around New

Table 3.2 Storm surge barriers built or modified after the Dutch Delta Works, 1987–2014

Year Completed/ Upgraded	Name	Location	Barrier Type	Hydraulic Head (m)	Construction Costs (in $ millions 2009)
1987	Thames barrier	London, United Kingdom	Sector gates	7.2	2043
1997	Maeslantkering	Rotterdam, Netherlands	Floating sector gate	5	925
2002	Ems Barrier	Ems, Germany	Sector gates	3.8	519
2010	SeaBrook barrier	New Orleans, United States	Vertical lifting gates/ Sector gates	4	162
2013	IHNC barrier	New Orleans, United States	Sector gates	4	730
2014	MOSE project	Venice, Italy	Flap gates	3	6596

Source: Modified from Hillen, M.M., Jonkman, S.N., Kanning, W., Kok, M., Geldenhuys, M.A., & Stive, M.J.F. (2010). *Coastal Defence Cost Estimates. Case Study of the Netherlands, New Orleans and Vietnam*. Delft, The Netherlands: Delft University of Technology (TU Delft).

Orleans and in southern Louisiana.[58] As an article in the *Harvard Design Magazine* lauds:

> The Thames and Eastern Scheldt barriers provide templates that can be applied to comparable situations ... The outer reaches of New York's waterways could be protected by a barrier on the model of the Eastern Scheldt in the Netherlands, connecting barrier beaches from Sandy Hook to Coney Island. If such a barrier were needed, it would be part of a system of seawalls that would protect the barrier beaches.[59]

One textbook even suggests that innovations such as the Eastern Scheldt Storm Surge Barrier become part of a package of "next generation infrastructure" to be considered by every city as our postindustrial society confronts climate change.[60]

4 Exposing the political economy of the Eastern Scheldt Storm Surge Barrier

Though certainly admirable in some of its scientific and technical attributes, and its ability to offer a template other countries have relied upon for their own climate-proofing, this part of the chapter shows that the Delta Works were also notable for becoming entangled with the four political economy attributes central to this book's thesis. We see enclosure at work, where the *Rijkswaterstaat* and other engineers and scientists were able to assert greater control over local water planning and policy. We see exclusion, where smaller construction firms, minority groups, NGOs, and environmental activists were excluded from the Delta Works construction or decision-making process. Due to the environmental nature of the project, we see encroachment – intruding on wetlands, destroying biodiversity, and creating significant environmental damage. We lastly see entrenchment, where the storm surge barrier in particular transferred the magnitude and frequency risks of flooding and the impacts of environmental pollution elsewhere, into already marginalized communities or degraded landscapes. This part of the chapter touches upon each in turn.

4.1 Enclosure

The process of enclosure affiliated with the Delta Works had two dimensions. Directly, the Delta Works enabled national planners, notably those at the *Rijkswaterstaat* to intensify their control over the local collective water organizations and councils that had been making decisions about water for upwards of 800 years. Indirectly, it facilitated the continued professionalization of scientists and engineers throughout the Delta Works process, enclosing upon roles previously held by local leaders and even laypeople. In this way, "the Delta Plan can be seen as a fundamental change in the balance between local and national water politics."[61]

Given the importance of flooding and water to the Dutch national identity, it might come as no surprise that the Dutch held a group of experts known as "dike masters" in high regard starting as early as the twelfth century. In that century, local water boards were created, and they were governed by councils elected by landowners who had voting rights in proportion to the size of their estates. The boards were responsible for community tasks including drainage, dike maintenance, and sluice management – and they had the authority of raising or levying taxes among other legislative and executive powers.[62] However, near the end of the eighteenth century, care for dikes and other hydraulic infrastructure shifted to the national government with the founding of the *Rijkswaterstaat*, an agency that was actually modeled on the *Corps des Ponts et Chaussees* (Corps of Bridges, Waters, and Forests) in France. The nineteenth century saw the *Rijkswaterstaat* become more professionalized, as civil engineers and other people with academic degrees succeeded in joining its ranks.

The Delta Works changed this balance of authority further, in complicated ways. In one sense, after the 1953 storm, the roles of both local boards and the *Rijkswaterstaat* were strengthened. Many local water boards were merged into stronger organizations with extended responsibilities. They became responsible, for instance, for all water treatment plants across the country. Moreover, many dikes of supra-regional importance were brought under their responsibility, after reinforcements were completed by the *Rijkswaterstaat*.

Nonetheless the authority and strength of the *Rijkswaterstaat* outpaced gains acceded to local water boards. As Disco notes, the Delta Works enabled the group's "professional monopoly" to extend "to the entire domain of hydraulic engineering and water management in the Netherlands, that is, to the provincial and even the local levels."[63] Control over some aspects of dike management was taken away from local water boards, and placed into the hands of national planners. This somewhat radical change in power was justified in part by the 1953 flood – a decentralized system of dike and water management was deemed culpable for the poor quality and maintenance of infrastructure prior to the storm. As sociologist Wiebe Bijker put it, "by the late 1950s, the faith in the *Rijkswaterstaat* was such that this centralization met with no resistance."[64] By the end of the decade, the *Rijkswaterstaat* – and especially the Department of Water Management within it – was completely in charge of sea defense, water quality, waterways, roads, and "all other aspects" related to flooding and coastal protection.[65]

In a less direct, but still salient manner, the Delta Works further strengthened the professionalization of water management in the Netherlands. They enabled scientists – in particular, civil engineers, hydraulic researchers, mathematicians, and modelers – to enclose upon tasks and roles formerly held by public servants at the local water boards. The establishment of a Delft Hydraulics Laboratory enabled engineers to refine and test key concepts in dredging, construction, and the production of mattresses.[66] Mathematics and advanced modeling also played a pivotal role in projects, with the storm surge barrier in particular requiring "the most complicated models" since they needed to assess the impacts of both salt and fresh water, and since the project was so large it necessitated analysis of wind and wave flumes for dikes and sluices.[67]

In sum, this expanded professional role of engineers, scientists, and technocrats at the *Rijkswaterstaat* was seen as instrumental in enabling the project to succeed.[68] Bijker has even gone so far as to argue that this ability to coordinate different actors – an "organizational innovation" in stakeholder management – was almost as significant as the technical and engineering innovations involved in actually building the physical parts of the Works.[69]

Still, following the completion of the Eastern Scheldt Storm Surge Barrier, Dutch planners, especially those at the *Rijkswaterstaat*, seemed cognizant that the concentration of water engineering expertise in one institution was a serious risk. For the next large storm surge barrier, the *Maeslantkering*, engineering companies competed to carry out the work; the *Rijkswaterstaat* remained on the sidelines. In that project, the role of the *Rijkswaterstaat* was reduced to procurement, contract management and supervision. In this manner, we see the *Rijkswaterstaat*'s relinquishment of power as an acknowledgment of the pitfalls of its early enclosure efforts.

4.2 Exclusion

The decision-making process behind the Delta Works was in some ways exclusionary, with two dimensions related to tendering and contracting, and – for the first few decades at least – discussions over design.

For instance, due to the size and capital intensity of the project, the *Rijkswaterstaat* presumed that contracting private construction companies, especially small and medium enterprises, would not work. The *Rijkswaterstaat* maintained that the Delta Works "were too difficult so as to be designed beforehand with enough detail to allow the normal bidding and contracting procedure" and that "distributing the various projects over many separate companies would severely hamper the 'Deltaschool' idea of accumulating knowledge and experience."[70] Thus, only a select number of construction companies were invited to participate, with the final consortium consisting of a mere eleven entities.[71] Smaller companies and dredging operators could only participate by joining as lower partners in the large consortia.

Another element of exclusion – most prevalent in the 1960s – relates to the decision-making and planning process. Though the *Rijkswaterstaat* featured "academically trained civil engineers" who were "quite competent in the design of hydraulic and transport infrastructure," they had "no training at all in the biological and ecological implications of their practice" and thus marginalized or ignored views challenging their expertise.[72] As Leemans explains, the *Rijkswaterstaat* considered themselves to have "a monopoly of knowledge" related to every aspect of water management and they operated in "a highly closed system."[73] At the outset of the Delta Works, the *Rijkswaterstaat* was what Disco called "opportunistically blind to the suppression of concerns other than safety."[74] To them, "no price seemed too high to ensure the statistical level of safety of one flood per 4,000 years."[75] As the *Rijkswaterstaat* itself wrote: "Safety compels the closure of the estuaries. This implies a certain loss of natural values which must be accepted."[76]

To be fair to the *Rijkswaterstaat,* at the time the Delta Works was proposed in the late 1950s, safety was seen as more important than environmental or social concerns, which were placed as secondary to national defense against flooding. Further, the *Rijkswaterstaat* did reach out to some stakeholders, and the organization was following the directives and laws of the time to the letter, and adhering to the principles of Dutch democracy. The *Rijkswaterstaat* formed an alliance with the chief engineers at the large construction companies mentioned above. That alliance also included an "advocacy coalition" of the Cabinet, the majority of Dutch Parliament, most senior representatives of municipalities – especially urban ones receiving the protection from the barrier – and some farming and agricultural organizations.[77] As political scientist Sander Meijerink explains:

> These parties strongly favored the implementation of large-scale coastal engineering works without addressing the negative ecological

impacts that these works entailed. The coalition attached great importance to safety, moderate importance to economic well-being, and no importance at all to preserving the environment.[78]

Similar to many other technocratic projects of the time, the coalition also depended upon exclusionary forms of information sharing and decision-making. For example, advocates of the Eastern Scheldt Storm Surge Barrier were reluctant to release information about designs and cost. For other departments within the government, "it was difficult if not impossible to obtain more detailed information and insight or to check the relevance of the arguments."[79]

As a result of this unified stance and strict control of information, other concerns and stakeholders were marginalized until the 1970s. In the 1960s, saltwater fishers and shellfish breeders started a protest movement to keep the Eastern Scheldt estuary "open" and the storm barrier either cancelled or substantially modified.[80] A second opposing coalition consisted of conservation scientists, representatives of environmental NGOs, and some representatives of ministries responsible for recreation; these actors formed or supported the creation of an Eastern Scheldt Study Group, an Action Group for an Open Eastern Scheldt, and a Committee for an Open Eastern Scheldt.[81] However, despite the protestations of these advocacy coalitions, "the water technocracy was fully in control, and little new knowledge, values and approaches were allowed to come in."[82] Those in favor of the Eastern Scheldt Storm Surge Barrier were able to insulate themselves from criticism and disregard the economic concerns of fishers and the environmental concerns of some scientists as frivolous. The often used "knockdown argument" was that "no new viewpoints justify new studies."[83]

Thankfully, things did change in the 1970s. By then, as Figure 3.5 reveals, a multitude of options were available to civil engineers. One of these – proposed in 1973 and later adopted in 1976 – was keeping the barrier "open" with a semi-permeable dam during calm weather, and closing it during storms. This option was more expensive but kept the Eastern Scheldt estuary connected to the North Sea and attempted to minimize negative impacts on fishers. As Disco notes, "The ingenious compromise 'spared both the cabbage and the goat,' as the Dutch saying goes, and was widely touted as a political and technological triumph."[84] Some have even gone so far as to argue that the decision to alter the design of the Eastern Scheldt Storm Surge Barrier to appease local economic and environmental interests amounted to a great "turning point" in Dutch water planning[85] and a major "transition in

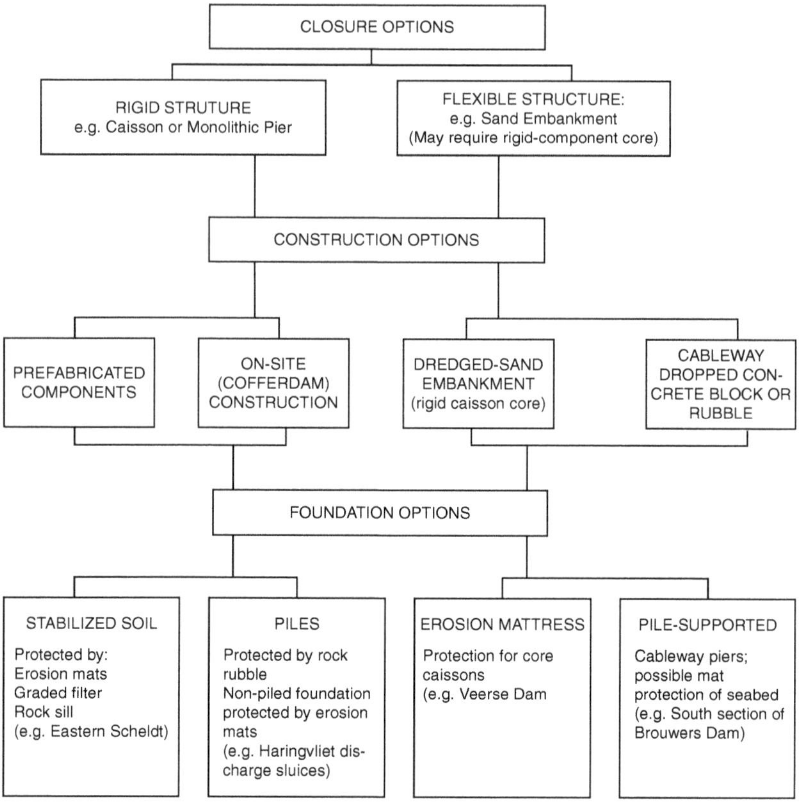

Figure 3.5 Closure, construction, and foundation options for the Eastern Scheldt Storm Surge Barrier

Source: Modified from Watson, I. & Finkl Jr., C.W. (1990). State of the art in storm-surge protection: The Netherlands Delta project. *Journal of Coastal Research*, 6(3), 739–764.

Dutch water management," a shift from a closed system of hierarchical control to a more participatory and decentralized *modus operandi*.[86] Nonetheless, even though the design of the Eastern Scheldt Storm Surge Barrier was altered to accommodate environmental concerns, the *Rijkswaterstaat* was able to maintain continuous control over the project. As Bijker concludes:

> When we take a closer look, we can see an argument to be made for continuity... Nobody questioned the basic safety goals of the Delta Law; ecological concerns were added to it... The Rijkswaterstaat recuperated after the slap in the face and regained control over the

> process, although for the contracting companies the Oosterschelde barrier remained one of the sweetest projects ever.[87]

Indeed, in June 1976 when the final plan was endorsed by Parliament, it was rushed through moments before summer recess with "hardly any possibility for hearing civil servants, experts, and relevant organizations;" it contained hastily compiled "technical, environmental, and especially financial information" that ended up "wrong in several respects."[88]

4.3 Encroachment

Perhaps the most comprehensive consequence of the Delta Works is encroachment: the project degraded fishing areas and eroded biodiversity and ecosystem vitality. Here, it depends upon what one uses as a basis for assessment. Compared to the original design, the modification to an "open" barrier was an environmental boon. Compared to leaving nature alone, however, the Delta Works and Eastern Scheldt Storm Surge Barrier in particular intruded upon and interfered with "an extremely complex hydrodynamic regime encompassing fast-flowing tidal channels, changing estuarine configurations, nonhomogeneous sub-bottom sediments, a prolific, but delicately-balanced ecosystem."[89] Such encroachment occurred through multiple drivers and levers with complex and long-lived negative impacts on multiple habitats.

To better understand these impacts, it is useful to start with some history about the Delta Works and specific details about the nature of the Eastern Scheldt ecosystem. In terms of history, when the Delta Works were endorsed by the government in 1957, environmental concerns hardly existed within government, the scientific community, or even the public. Priorities in the late 1940s and 1950s centered on reconstruction after World War II; priorities in the mid-to-late 1950s centered on safety, as previously described. At that time, no public organization was in charge of the environment. The *Rijkswaterstaat* was the first governmental agency to create a strong environmental study department in 1972. At that time, environmental awareness had gained an important place in political and social organizations. When the government asked the *Rijkswaterstaat* to come up with alternative solutions to a closed barrier design, the organization studied the matter intently, and also enlisted the help of the RAND Corporation.[90]

However, given the unique nature of the Eastern Scheldt estuary, these early efforts were not able to effectively curtail environmental damage. Before the Eastern Scheldt Storm Surge Barrier, biologists catalogued approximately 850 different species of plants and animals residing there.[91] During the summer months, about 150,000 water birds were usually present per every square half kilometer of the estuary, giving it incredible "ornithological value." As one study explained in 1984, before the storm surge barrier was completed:

> The Eastern Scheldt is strikingly rich in bird species, especially waders and ducks. Of great importance in this connection is the rich supply of food associated with the large quantity of benthic organisms in the intertidal areas. In addition the Eastern Scheldt and its surroundings are a relatively peaceful area in which birds can find refuge at high water ... Eastern Scheldt has a vital function in nature conservation: this is because of its scale and spaciousness, its tidal dynamics, its tidal landscapes – such as the salt marshes – which are becoming ever rarer.[92]

The Eastern Scheldt was so prized for its biodiversity that it was one of only four areas in the Netherlands suggested for protection under the 1971 Convention on Wetlands (the "Ramsar Convention").[93]

Unfortunately, compared to a completely open, unaltered habitat, the Eastern Scheldt Storm Surge Barrier negatively impacted the stability of this ecosystem in many ways.[94] It altered land morphology and reduced the tidal area of the estuary from 450 square kilometers to 330 square kilometers. It reduced water exchanges with the North Sea by 28 percent and reduced freshwater loads by 64 percent. It suffered from "sand demand" or "sand hunger," meaning it lessened the speed of currents and altered the velocity and timing of tidal flows so that they no longer transported sediment through gullies and into tidal areas – essentially, a slow erosion of land occurred. In other words, the barrier caused large parts of sandbars to eventually disappear. It changed mean tidal amplitude, which dropped from 3.36 meters to 2.7 meters. It converted dried land into brackish land, induced subsidence, and altered water quality – especially the salinity and chlorinity of water. Due to a reduction in total water volume, it changed the entire temperature of the ecosystem. It decreased nutrient loads across the entire ecosystem by 44 to 82 percent.

These changes, in the aggregate, resulted in a new "unnatural ecosystem."[95] Early on, scientists warned that the barrier would convert a tidal area rich in biodiversity, especially avian fauna, into a stagnant lake with greatly reduced water quality.[96] In a late 1970s assessment, Parma determined that the barrier would exert "large changes in the chemical and physical environment."[97] Wetsteyn and Krompkamp concluded that:

> The construction of the Eastern Scheldt Barrier and associated dams will have a great impact upon two ecological habitats of the estuary. The salt marsh area will be reduced by nearly 60 percent. Tidal flat area will decrease by 45 percent, which will have major repercussions for the present role of the Eastern Scheldt as a major foraging station for migrant and wintering waders.[98]

Such projections do not seem hyperbolic in hindsight, given that studies after the completion of the barrier noted a combination of premature mortality and morbidity among mussels,[99] the collapse of marsh vegetation,[100] unusually high abnormalities among sea snails,[101] and even the death of larger mammals such as harbor porpoises, with one team concluding that "mortality is exceptionally high in the Eastern Scheldt."[102] Birds have been the most severely impacted, due to shifts in habitat, a reduction of foraging area, and losses in total carrying capacity. One study estimated population losses for all birds by 18 percent and some bird species as much as 30 percent, as Table 3.3 indicates.[103] Another research team followed animals living within the estuary at three monitoring stations over the course of a year and (depressingly) noted that within six months, many of the former intertidal flats had

Table 3.3 Numbers of avian waders in winter after construction of the Eastern Scheldt Storm Surge Barrier

	N (winter)	**Change (%)**
Haematopus ostralegus (Oystercatcher)	75,656	(–23)
Recurvirostra avosetta (Avocet)	222	(+45)
Charadrius hiaticula (Ringed Plover)	57	(0)
Pluvialis squatarola (Grey Plover)	5,892	(0)
Calidris canutus (Knot)	12,523	(–7)
Calidris alpina (Dunlin)	37,295	(–30)
Limosa lapponica (Bar-tailed Godwit)	6,647	(–9)
Numenius arquata (Curlew)	7,262	(–15)
Tringa erythropus (Spotted Redshank)	23	(0)
Tringa totanus (Redshank)	1,492	(–30)
Arenaria interpres (Turnstone)	1,077	(–20)
All birds	148,146	(–18)

Source: Modified from Smaal & Nienhuis (1992).

become permanently submerged with stagnant saltwater, meaning "most species died off."[104] As one report declared when reviewing the environmental impacts of the project: "it has become clear that the Eastern Scheldt is a basin that has been shaped strongly by a multitude of human interventions... It will take in the order of centuries before the morphological effects of these interventions will have leveled out."[105]

Nor are the environmental consequences of the Delta Works limited to the environmental domain. One environmental group warns that unexpected erosion following construction of the Eastern Scheldt Storm Surge Barrier will result in substantial loss of economically valuable land, with some 9,500 hectares of intertidal land lost by 2100.[106] The group estimates that the expected land lost will be greater than the losses expected due to sea level rise, the implicit argument being that it would have been better to do nothing. Perhaps more alarmingly, the group also argues that as intertidal areas in front of the barrier recede, environmental buffers will be eroded and flood risks will increase, rendering additional investments necessary to protect lives, land, and property. The group estimated that these costs could rise as high as €260 million depending on rates of sea level rise.

Intuitively, the Eastern Scheldt was not the only piece of the Delta Works to impact the environment. Parts of the Works not only degraded the Eastern Scheldt estuary, they also concentrated pollution in other areas. Those along the waterways of the Rhine and Maas rivers transformed transparent water lakes into huge sinks.[107] The River Scheldt itself, which had become a drain for various types of untreated waste (industrial, agricultural, and household), was for all intents and purposes plugged; its ability to disseminate such pollutants into the North Sea diminished.[108]

4.4 Entrenchment

The entrenchment involved in the Dutch Delta Works relates to a socialization of the risks of flooding to other parts of the Netherlands. The Works precipitated a false sense of security and lulled Dutch planners into thinking that they had eliminated flooding risks. It is telling that in 1993 and 1995 water levels in the Rhine and Meuse rivers rose to extreme heights, and in 1995, 200,000 people had to be evacuated.[109] Van der Brugge, Rotmans, and Loorbach warn that in some ways the Eastern Scheldt Storm Surge Barrier has only entrenched selective vulnerability:

> Now it becomes increasingly clear that this "pumping-drainage-dike raising" strategy has not resulted in a sustainable water system as many calamities have occurred. The incremental reduction of space

> for the natural water system (polders and usage of river forelands) and changes in the water system itself (reduction of the natural riverbed and hydrological conditions) has manifested itself in social damage (reduced safety), financial damage (floods, droughts, dike breaks) and ecological damage (drought, loss of water quality).[110]

Another major report warned in 2000 that the Netherlands and its water management strategy was "not sufficiently prepared to meet the challenges of climate change effects in the next century."[111] That is, it was not sufficiently protecting all of the population as it had promised.

It is telling that following the Delta Works, the entire approach to water management and flood risks has changed in the Netherlands. Post the 1995 flood, a new strategy arose centered on "living with water" or making "room for the river" rather than further raising dikes and hardening infrastructure. The concept was that instead of protecting all areas of the Netherlands, occasional flooding would be "accommodated and carefully managed in specific designated areas."[112] Interestingly, it is peripheral communities that have been selected for this "controlled inundation" of water, with them coming to be described as "calamity polders" – parts of Dutch society kind of written off, and left unprotected from flooding.[113] As one study puts it, "these [calamity] polders were

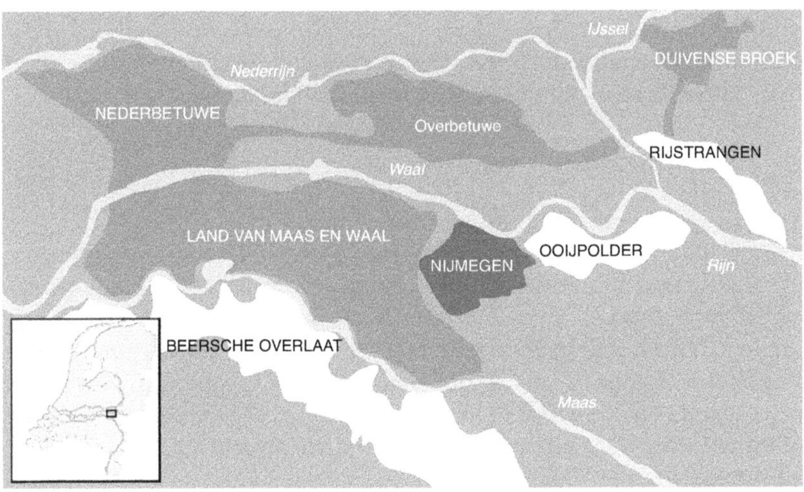

Figure 3.6 Three planned Dutch "calamity polders"

Note: The three calamity polders appear in white.

Source: Modified from Roth, D. & Warner, J. (2007). Flood risk, uncertainty and changing river protection policy in the Netherlands: The case of "calamity polders." *Tijdschrift voor Economische en Sociale Geografie*, 98(4), 519–525.

located in areas with a low-population density to take the pressure off the highly urbanized, economically more important and more densely populated western part of the Netherlands."[114] Indeed, Figure 3.6 depicts the three proposed "calamity polders" of Rijnstrangen, Ooijpolder, and Beersche Overlaat – all three rural, relatively marginalized areas of the Netherlands.

Dutch planners and civil engineers are also becoming more aware that flooding and storm surge are not the only predominant climate risks, and that by focusing exclusively on them, they might have missed opportunities to build resilience in other sectors. For instance, in 2003, a European heatwave resulted in 500 deaths of mostly elderly people. That same year, low freshwater discharges from the Rhine resulted in saltwater intrusion of water supplies and shortages for agriculture and horticulture. Higher summer temperatures precipitated serious shortages of water for the thermoelectric cooling cycles of power plants, with somewhat severe implications for industrial output.[115] Though it is debatable which types of risks will be most severe, each necessitates its own type of (expensive) intervention to build adaptive capacity.

Most seriously, in 2008, the Delta Commission announced that if sea levels rise greater than 1.0 meters, which is possible given current projections, then the entire Eastern Scheldt storm surge barrier "cannot operate safely...and preferably has to be removed."[116] This means the project could in one sense backfire, and necessitate an expensive, time-consuming, intricate decommissioning process – which would cost at least $1.6 billion – and offset all of its potential benefits.

5 Uncomfortable conclusions and implications

This part of the chapter presents four conclusions.

First, like our other cases, the processes of enclosure, exclusion, encroachment, and entrenchment are multi-scalar. As Table 3.4 indicates, even though members of the *Rijkswaterstaat* likely saw themselves as loyally carrying out the tasks requested by the government or the public, macro drivers at the national scale included the awarding of contracts to large Dutch construction champions, and vesting the *Rijkswaterstaat* with supreme control over the project and enabling it to withstand alternative designs and criticism. Meso drivers at the regional scale included a deterioration of water quality and degradation of the precious ecosystem found within the Eastern Scheldt. Micro drivers related to the intensification of pollution in areas adjacent to the barrier and entrenchment of flood risk in so-called "calamity polders."

Table 3.4 The multi-scalar nature of the political economy of the Dutch Delta Works

Scale	**Political Economy Predominantly Affects**	**Explanation**
Macro: National actors (the *Rijkswaterstaat*, Dutch parliament, large national corporations)	Disbursement of contracts for construction, access to decision-making and even information	Construction contracts get directed only to large national firms, the *Rijkswaterstaat* exerts a command and control management strategy that limits access to information and stifles political debate
Meso: the Eastern Scheldt Estuary	Ecosystem vitality and the abundance of biodiversity found within the Eastern Scheldt	Marsh vegetation, mussels, snails, harbor porpoises and especially birds see their habitat and spawning ground severely degraded by the project
Micro: Riverine communities, calamity polders	Concentrations of pollution, flooding risk, vulnerability to future climate disasters	The storm surge barrier shifts concentrations of toxins to the Western Scheldt and other rivers and also entrenches flood risk among some peripheral communities

Source: Authors.

Second, and for the good, Dutch planners, especially those at the *Rijkswaterstaat*, seem aware of some of their earlier mistakes; they have made meaningful attempts to learn from the past. The *Rijkswaterstaat* acknowledged, perhaps as a major lesson following the completion of the Eastern Scheldt Storm Surge Barrier, that their role should be limited to procurement, contract management and supervision, and that a less centralized, more cost-conscious strategy was warranted. Their style of management became more inclusive, open, and competitive.

Third, even though it was a child of the 1950s and 1960s, the Dutch Delta Works attained a sort of autonomy that made it difficult to alter the stated path of the collective projects. Scholars writing in the energy and climate fields have termed this "path dependence,"[117] "lock-in,"[118] and socio-technical "momentum."[119] These are various terms that describe the difficulty of altering a large-scale technological system once it gets going. In the case of the Eastern Scheldt Storm Surge Barrier, for instance, even in the 1970s, when environmental awareness was high across the Netherlands, the infusion of sustainability ideals and attitudes was incomplete. As a result, the final plan for the project rushed through Parliament with little debate and resulted in severe negative environmental consequences.

Fourth, and perhaps most seriously, the situation with the Delta Works reminds us that floods and extreme events are difficult to predict. Despite the $4.7 billion spent on the Eastern Scheldt Storm Surge Barrier and the almost $10.3 billion funneled into the Delta Works, those projects did not eliminate the risk of flooding to one in 4,000 years for the Netherlands. After decades of careful modeling, planning, and construction, the centerpiece of the Delta Works – the Eastern Scheldt Storm Surge Barrier – might need to be removed, at the expense of $1.6 billion, if sea levels rise beyond 1.0 meters. This reminds us that building resilience requires not only engineering components such as storm surge barriers with semi-permeable dams, but also an attention to the social aspects of vulnerability, financial aspects of costs, and political aspects related to "acceptable" losses.[120]

Such a lesson becomes all the more salient when one considers that the storm surge barrier has become a template for climate resilient infrastructure around the world, one that has been exported to other countries. A massive public works project was launched in the 1950s to tackle what was then perceived to be the greatest threat to Dutch national security – flooding from severe storms – only to become largely outdated under newer, more sophisticated projections of climate change and sea level rise. Perhaps given the uncertainty involved in anticipating and addressing future climate impacts, the real lesson is that complex, capital intensive, climate-proofing infrastructure projects such as the Delta Works address some risks only by ignoring others.

4
Bloated Bodies: The Political Economy of Hurricane Katrina Recovery

1 Introduction

On the morning of August 29, 2005, Hurricane Katrina made landfall in the United States, hitting the coast of Louisiana with wind speeds of 125 miles per hour near Buras-Triumph.[1] It was not the strongest of storms in terms of wind speeds, central pressures, or intensity – it was only a Category 3 hurricane – but its particular location along the Gulf Coast, and its abundant amount of rainfall and storm surge made it "the most devastating and costly hurricane in US history."[2] At least 1,800 people died in the hurricane and its subsequent floods, and total cleanup and repair costs have been estimated at $80 billion to $400 billion.[3] Although the storm's wrath was felt from central Florida to Southern Texas, most of the damage occurred in New Orleans, Louisiana, after its levee system failed and about 80 percent of the city was flooded with stagnant water for many weeks.[4] That catastrophe was so stark that one expert has called it "the worst civil engineering disaster" in American history.[5]

What in the world happened? How could a hurricane predicted more than a week earlier result in so much damage? As this chapter shows, the four political economy processes of enclosure, exclusion, encroachment, and entrenchment both worsened the impacts of the storm, and led to their own set of pernicious social, environmental, and economic consequences. As sociologists Robert D. Bullard and Beverly Wright explain, Katrina offers a sobering example of discrimination and inequality affecting multiple stages of disaster recovery, spanning "preparedness, communication, physical impacts, psychological impacts, emergency response, clean-up, recovery, and reconstruction."[6] To make this case, focusing primarily on New Orleans and Louisiana but also (when warranted) other states, this chapter first traces the historical development of the region's political economy before detailing the genesis and impacts of the storm. It then exposes the political economy factors at play in recovery efforts before offering five broader conclusions. To be sure, the interconnection of these four political economy processes did not happen in a vacuum; the political history of the area paved the way for the Katrina crisis.

2 Southern political economy and a prelude to disaster

To gain insight into the complexities of relief and reconstruction following Hurricane Katrina, it is necessary to first delve into the historical political economy of the South and New Orleans, a past that left the region eminently vulnerable to disasters. As we will see, "these historic developments have coalesced to produce a peripheral region characterized by deep and complex relations of racial and class division."[7]

Political economy factors played a formative role in the development of New Orleans and the Southern economies all the way back to their founding. One culprit certainly dates to the early 1700s, when the United States was still a British Colony, and the Crown's policy explicitly encouraged wealthy Anglicans who could afford slaves to migrate to the South, and more industrious religious refugees without property to migrate to the northern colonies. This policy led to a bifurcation of the North and South, setting in motion two distinct socioeconomic systems that culminated in a Civil War more than a century later.[8]

The history of New Orleans dates to at least 1718, when Jean-Baptiste Le Moyne de Bienville, a French Canadian, as well as governor of the French Louisiana territories, left what is now Alabama with a small group of people to establish a city on the banks of the Mississippi, located 90 miles from the Gulf of Mexico, named in honor of the Duke of Orleans. *La Nouvelle Orleans* was originally established as a military outpost but quickly became a trading hub and then administrative center for French land holdings in the new state of Louisiana.[9] It was an integral part of the slave trade, with one historian estimating that New Orleans was "the largest slave market in the United States," with two-thirds of all slaves brought to the Southern states passing through the New Orleans port; these slaves represented roughly $500 million in historical "property value."[10]

This uneven (and highly racist, slave-dependent) development became even more hierarchical and regimented during the mid-to-late 1800s, as the growth of transcontinental railroads in the United States linked southern agriculturalists with northern sources of capital. Although these linkages initially seemed net beneficial to both North and South – northern financiers were able to access markets that gave them great returns on investment, southern farmers received much-needed revenue – they eventually made the Deep South dependent on the North.

This dependence only became starker as the growing industrial centers in the Midwest and West marginalized the importance of the Mississippi River as a source of trade.[11] Such dependence was certainly one (of many) causal factors in the American Civil War from 1861 to 1865. After the Civil War ended, the population of racial minorities in the Deep South swelled, with the poorest African Americans living wherever they could, acquiring cheap property, when available, that nobody else wanted. Thus began a cycle of displacement where poor and minority communities continually had to live on the most vulnerable or "worst" land.[12]

For a brief period of time, the trend toward racial disempowerment and classist development faltered. In 1872, the first African American governor of a US state was elected, and the city of New Orleans operated a progressive, racially integrated public school system.[13] However, these gains were short-lived, and after a Reconstruction Period ended in Louisiana in 1877, white supremacists regained control of the legislature; they forced racial segregation of the schools and further disenfranchisement of African Americans.[14]

The period of 1900 to 1960 saw confidence in technology paired with a disregard for social circumstances; it only worsened the vulnerability of minority and poor groups in the South. It was during this time that "the safe development paradox" emerged.[15] Industrialization and enhanced drainage techniques piloted and perfected in the 1900s and 1910s gave planners supreme faith that technology and infrastructure could stall the impacts of any disaster. This facilitated a pattern of urbanization integrated with levees and protective infrastructure – people lived literally right behind or in some cases on top of canals, polders, dikes, and so on. Kates calls this a "paradox" because ultimately the belief that the levees were safe led people to move closer to them, undermining their ability to withstand the impacts of severe storms. A severe hurricane in 1915 damaged some 25,000 buildings, overwhelming the levees. A 1947 hurricane storm surge overtopped lakefront levees in suburban neighborhoods. Hurricane Betsy flooded almost all of the Lower Ninth Ward in New Orleans in 1965, displaced 300,000 residents, and destroyed 27,000 homes.[16]

Because vulnerability is about exposure to disaster and resilience to it, it was predictable that the poor and African Americans would suffer the bulk of the impacts of these hurricanes. As Kates and his colleagues surmise:

> In its 288-year history, New Orleans has had 27 major river or hurricane-induced disasters at a rate of one about every 11 years. A pattern of three responses runs through that history. After each event, the city rebuilt and often expanded, small differences in elevation determined the location of the well-to-do and the poor, and levees were rebuilt and often raised. Inequity in the location of neighborhoods and in the distribution of flooding burdens also appears early.[17]

The authors note that this pattern of response began in the 1800s and continued across all instances examined. Thus, powerful actors have long responded to natural disasters in New Orleans and along the southeastern coast by furthering their own interests.[18]

Unfortunately, three more recent trends in geology, industrialization, and political neglect had only heightened the vulnerability of New Orleans and the Gulf Coast: loss of elevation and a rising sea level; a weakening of wetlands and environmental buffers to severe storms; and lack of proper maintenance and upgrades to the levee system. The first trend has been a gradual loss of elevation within New Orleans due to higher than expected rates of subsidence coupled with unanticipated increases in sea level rise.[19] Wright and Bullard note that, for all intents and purposes, modern day New Orleans is situated in "a bowl nearly surrounded by water," bordered by the Mississippi River on one end, Lake Pontchartrain on the other, and entirely dependent on an intricate system of levees and canals to stop flooding.[20]

A second trend has been the erosion of an environmental buffer. The Gulf Coast region is home to more than 4,000 offshore oil wells, which produce more oil and gas than the country imports from Saudi Arabia.[21] As McDougall surmised, "Refineries, chemical processing plants, and the nation's Strategic Petroleum Reserve round out a massive energy infrastructure, with New Orleans at the epicenter."[22] These facilities have wreaked havoc on environmental systems that ordinarily would provide natural protection against hurricanes. For instance, offshore oil drilling, pipelines, highway projects, and industrial development have cumulatively destroyed 2,000 of the original 7,000 square miles of coastal marsh and swampland around the coastal delta of the Mississippi River.[23] Since 1940, roughly one million acres of coastal wetlands have been converted to open water as a result of human development. As one study explains, "the wetlands and protective barrier islands that would dampen storm surges and waves during hurricanes have been lost, therefore increasing the risk of flood disaster."[24] Resilience has been further weakened by chronic, low-level releases of many toxic chemicals through the discharge and seafloor accumulation of drilling muds and cuttings, as well as the continual release of hydrocarbons and polluted water around production platforms.[25] From 1980 to 1999, for instance, more than three million gallons of oil were released by accident or intentional discharge into the Gulf of Mexico, with severe impacts on coastal wildlife and the integrity of wetlands.[26]

A third trend has been political neglect of coastal protection infrastructure such as levees and canals, with privatization of many formerly government run services and substantial reductions in public budgets.[27] Before Katrina, the federal government reduced funding for the Army Corps of Engineers – those responsible for the levee system – by 80 percent, making it impossible for necessary pumping and levee improvements.[28] Similarly, federal and state governments continued to pour money into canals instead of restoring wetlands. For instance, the Mississippi River-Gulf Outlet canal, a primary cause of the flooding in Katrina, is a deep draft seaway channel that extends about 76 miles; it is intended to shorten the distance for ships traversing the eastern shipping lanes of the Gulf to New Orleans. Despite the fact that it only handles about one ship per day, or three percent of New Orleans port cargo traffic, combined federal and state subsidies to the canal amounted to about $20,000 per ship – an amount of money so significant it left an ecosystem restoration plan known as *Coast 2050* unfunded. [29]Furthermore, those in favor of closing the Mississippi River-Gulf Outlet cautioned that due to the deterioration of wetlands in the un-leveed stretches on the south and east parts of the canal, it represented "a shotgun pointed straight at New Orleans."[30]

In sum, the Gulf Coast region, and New Orleans in particular, was ripe for disaster well before Katrina made landfall. As Weber put it, "the extreme inequality in these Southern states reflects a legacy of government/elite/corporate alliances that promoted slavery and the plantation system, post-slavery agricultural peonage, the convict-lease system, emerging agribusiness, and more recently the non-union, low-wage, and internationally driven industrial/retail sector."[31] As we will see, the political economy of this alliance determined how disaster relief resources were allocated (setting the rules), who got them (winning the contest), and how things were depicted in the media (spinning the story).[32] It is within the context of this complex physical geography and racial history that the City of New Orleans and the Gulf Coast states were faced with their greatest natural disaster in the summer of 2005.

3 The history of Katrina and impacts of the storm

As the timeline summarized in Table 4.1 depicts, Hurricane Katrina formed as a tropical depression southeast of Nassau, Bahamas, on August 23, 2005, crossing into Florida as a mere Category 1 Hurricane before intensifying over the warmer waters of the Gulf of Mexico. It strengthened to a Category 5 Hurricane with 175 mile per hour winds while out at sea, but became more moderate as it approached land in southeast Louisiana; it was classified as a Category 3 Hurricane the morning of Monday, August 29. When the storm struck New Orleans early in the morning, its winds had dropped below 130 miles per hour, but within three hours the storm breached the floodwall of the Lower Ninth Ward.[33]

Table 4.1 Chronological timeline of Hurricane Katrina, August 23 to September 5, 2005

Date	Event
Tuesday, 23rd August	Hurricane Katrina forms in the Bahamas Hurricane warnings are posted in Southern Florida
Thursday, 25th August	Katrina becomes a Category 1 Hurricane and makes landfall in Southern Florida with sustained winds above 80 miles per hour
Friday, 26th August	Katrina travels across Gulf of Mexico and achieves major Hurricane Status Louisiana Governor Kathleen Blanco declares a state of emergency Gulf Coast states including Louisiana begin requesting National Guard support and other federal assistance
Saturday, 27th August	Hurricane warnings issued for states of Alabama, Mississippi and Louisiana Failure to execute the City of New Orleans evacuation plan which relies on the use of school buses for evacuation purposes City rejects offer from Amtrak to evacuate citizens by train. Last train departs New Orleans without passengers At the request of Louisiana Governor Kathleen Blanco, President Bush declared a federal state of emergency in Louisiana, specifically authorizing FEMA to coordinate all disaster relief and to identify, mobilize, and provide at its discretion, equipment, and resources necessary to alleviate the impacts of the emergency

Continued

Table 4.1 Continued

Date	Event
Sunday, 28th August	Hurricane Katrina reaches Category 4 intensity with 145 mph winds It later becomes Category 5 storm, with maximum sustained winds of 175 mph In a press conference New Orleans mayor Nagin issues the first ever mandatory evacuation order for the entire city The Louisiana Superdome is opened as a refuge for residents who are unable to obtain transportation out of the city. The Louisiana National Guard delivers three truckloads of water and seven truckloads of food rations to the Superdome, enough to supply 15,000 people for three days
Monday, 29th August	Storm causes breaches in the levee system that protects the city of New Orleans from flooding
Tuesday, 30th August	Levee breaches cannot be plugged. Buses from across Louisiana are commandeered, and participate in the evacuation of 15,000 people to the Superdome 80% of New Orleans is under water as a result of breaches in the levee system
Thursday, 1st September	National Guardsmen accompanied by buses (475 in all) and supply trucks begin to arrive at the Superdome. Repair of levee breaches begins. President Bush signs a $10.5 billion relief package passed by the US Congress. The President visits the hurricane-battered Gulf Coast, but receives criticism for delaying his visit.
Friday, 2nd September	18,678 Army National Guard and 2,464 Air National Guard units are deployed
Sunday, 4th September	The evacuation of the Superdome is completed
Monday, 5th September	The 17th Street Canal levee breach is closed with truckloads of rock and sandbags. The canal reopens so that it can be used to pump water out of city

Source: Compiled and modified from National Climate Data Center (2005). *Hurricane Katrina*. Washington, DC: National Oceanic and Atmospheric Administration, Available at: http://www.ncdc.noaa.gov/extremeevents/specialreports/Hurricane-Katrina.pdf; Driesen, D.M. et al. (2005). An unnatural disaster: The aftermath of Hurricane Katrina. Member Scholars of the Center for Progressive Reform; Baker, C.R. (2014). Breakdowns of accountability in the face of natural disasters: The case of Hurricane Katrina. *Critical Perspectives on Accounting*, 25(7), 620–632.

In the proceeding hours, multiple other levees and flood barriers broke, including an estimated 90 percent of the Mississippi River-Gulf Outlet canal in St. Bernard Parish,[34] ultimately covering more than 80 percent of the city in water as high as 20 feet – as Figure 4.1 reveals. Then, quite

Figure 4.1 Hurricane Katrina flooding of Interstate 10 and Interstate 610 in New Orleans, August 29, 2005

Source: US Coast Guard.

simply, mayhem ensued, with close to one million residents along the coast calling for help and many, when it did not come, forced to take refuge in the Louisiana Superdome, the Convention Center, hospitals, and on the roofs of their homes.[35]

Tragically, a series of human errors and bungled efforts only exacerbated the impacts of the storm. To name a few, on Saturday, August 27:

- The Mayor of New Orleans, Ray Nagin, ignored the city's established disaster plan, which required that public school buses evacuate residents unable to depart on their own. Because the city never deployed the buses, they ended up being decimated by flooding;[36]
- A national government passenger train service ran a special train to move federal equipment out of the city. The train had room for several hundred passengers, and Amtrak offered these spaces to the city, but the city declined them.[37]

On Sunday, August 28:

- Despite adequate warning 56 hours before landfall, orders for mandatory evacuation of the most vulnerable areas – those inhabited disproportionately by minorities – were delayed until 19 hours before landfall.[38]

On Monday, August 29, the day the storm hits Louisiana in the early morning:

- The Federal Emergency Management Agency (FEMA), tasked with overseeing national disaster relief efforts, sent hundreds of volunteers for two days of training classes in Atlanta, Georgia, rather than transporting them immediately in the disaster area;[39]
- The Transportation Security Administration (TSA) insisted that all evacuees and their luggage be screened before aircraft could depart, but x-ray machines were inoperable due to lack of electricity – causing an eight hour delay as hand-screening teams were flown in. Further delays occurred while the TSA secured air marshals to ride the planes;[40]
- Michael Brown, the head of FEMA, urged all fire and emergency services departments not to respond to counties and states affected by Hurricane Katrina until mutual aid agreements could be completed, he also prevented the US Forest Service from using water-tanker aircraft to extinguish riverfront fires.[41]

On Tuesday, August 30:

- The Louisiana governor's office requested helicopter assistance from the Pentagon for aircraft idling at Fort Polk, but received permission late in the afternoon when the pilots had already gone over their allotted time – they spent all day on the tarmac waiting for approval;[42]
- FEMA refused to approve volunteer rescue efforts offered by hundreds of small boats and watercraft and rejected offers of help from the Florida Airboat Association;[43]
- FEMA interfered with and prohibited private efforts to evacuate victims including stopping a hotel's plans to hire ten buses to carry 500 guests to higher ground, preventing Wal-Mart trucks loaded with water from entering the city, and prohibiting more than 50 civilian aircraft from responding to evacuation requests from hospitals.[44]

On Monday, September 5:

- FEMA realized that its original evacuation scheme had unwittingly separated families by hundreds of miles in different states, including 5,000 children separated from their parents;[45]
- More than 7,000 single-family houses owned by the Department of Veteran Affairs were to be used for evacuees but were rejected by FEMA, and remained vacant for three more months because of paperwork problems.[46]

On Thursday, September 8:

- A temporary morgue was finally established by FEMA. FEMA had refused for ten days to search for the dead and a FEMA spokesperson said that "finding and collecting these bodies were below the station" of its disaster response teams. Bloated bodies lay in the streets, rivers, and canals for days.[47] Figure 4.2 shows one such body near the Superdome taken in September, days after the storm hit.

Any single one of these incidents would have meaningfully worsened the impact of the storm. Collectively, they amounted to what *The New York Times* called "one of the most extraordinary... and stupefying bureaucratic bungles in modern history."[48]

Because of this mutually destructive confluence of negative natural and sociopolitical factors, some of the specific statistics about the storm

Figure 4.2 A bloated body in the floodwaters near the New Orleans Superdome, September 2005
Source: US National Guard.

are startling, and help put its broader consequences into perspective. The Hurricane impacted a total of 90,000 square miles,[49] an area roughly the size of the United Kingdom. The storm destroyed, in part or whole, 300,000 homes, damaged 150,000 businesses, and caused $80 to $400 billion in economic losses, making it the costliest disaster in US history.[50] More than 110,000 of 180,000 homes in New Orleans were flooded, and half sat for days or even weeks in more than six feet of water. After Katrina, 350,000 automobiles had to be drained of oil and gasoline and then "recycled," in addition to 60,000 boats, 300,000 underground fuel tanks, and 42,000 tons of hazardous waste.[51] Katrina displaced at least 1.5 million people and saw a similar number of disaster relief applications – the largest ever recorded – leading to "one of the largest disaster diasporas" in the history of the country.[52] The hurricane directly resulted in the deaths of almost two thousand people, along with hundreds more from "indirect causes" such as the evacuation, suicide, and drug overdoses.[53] Experts predict that while the storm lasted only a few days, and immediate disaster recovery lasted about 11 months, reconstruction will take eight to eleven *years*.[54] As Elliot and Pais argued, "In US history, there are no precedents for this degree of sudden devastation and outmigration from a major urban region."[55]

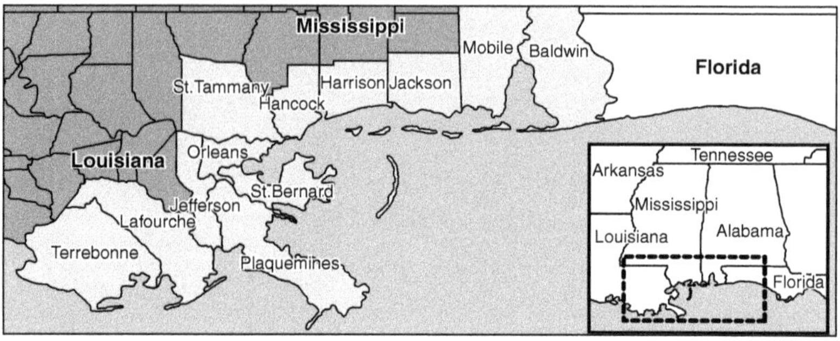

Figure 4.3 Communities affected by Hurricane Katrina flood and storm surges
Source: Modified from Cutter, S.L., et al. (2006). The long road home: Race, class, and recovery from Hurricane Katrina. *Environment*, 48(2), 8–20.

Furthermore, even though New Orleans and Louisiana received the brunt of the storm's impact, and the most media attention, Figure 4.3 shows that they were not the only areas affected. The worst property damage (in absolute terms) occurred in beachfront and coastal towns across Mississippi,[56] where some communities saw 90 percent of their infrastructure destroyed. Across these parts, boats and barges crashed into buildings and pushed cars and houses inland by as far as four miles; water from storm surges as high as 30 feet rushed in as far as 12 miles (19 kilometers).[57]

There is more, however, to Katrina and its disaster relief efforts than bureaucratic bungling, economic impact, human turmoil, and environmental destruction. As we will see in the next section, the disaster also presented an eminently attractive opportunity for the political economy processes aggravating inequality to thrive.

4 Exposing the political economy of Hurricane Katrina disaster relief

The central thesis of our book is that adaptation projects, including the disaster recovery efforts following Hurricane Katrina, succumb to four types of structural flaws: enclosure (capturing public resources), exclusion (limiting access to power), encroachment (degrading the environment), and entrenchment (worsening inequality). We find strong confirmation of each of these four processes in Katrina, as this part of the chapter demonstrates.

4.1 Enclosure

Enclosure refers to when an adaptation project shifts public assets into private control, or expands the role of private actors into the public sphere. In terms of Katrina, we see four different yet interconnected types of enclosure: the militarization of disaster relief by agencies such as the National Guard; expanded corporate control and profiteering over previously public services; the capture of public housing by private actors; and the takeover of public schools.

First, state and even national inaction created an opportune chance for certain actors to expand their roles into new areas. The most obvious example is the securitization of disaster relief and the strong role of the US National Guard and other active military forces from the Army, Marines, Navy, and Coast Guard. Though their involvement brought a degree of stability, certainly reduced loss of life, and mitigated acts of looting, rioting, and violence, it witnessed more than 50,000 military personnel taking on non-military responsibilities such as repairing breached levees, distributing food, offering medical care, and removing debris.[58] These efforts were in addition to "standard" military tasks such as flying search and rescue flights, maintaining crowd control, and policing. The non-military tasks enclosed – perhaps justifiably – on roles usually reserved for the Louisiana Recovery Authority, the Mayor's Office, and the Army Corps of Engineers. Some have argued that the militarization of relief during Katrina marked a turning point for the country in how it addressed future disasters; the US now pursues a more masculine, security-oriented approach through the Department of Homeland Security and FEMA.[59]

Second, in the immediate aftermath of Katrina, business as usual was suspended in terms of government rules for bidding on reconstruction projects and contracting. With disaster imminent, the thinking went, there was little time to solicit and weigh bids. Unfortunately, in practice this meant that the bulk of the contracts went to a bevy of large corporations, not small and medium enterprises. For instance, Katrina affected at least 60,000 black-owned businesses across the Gulf, which generated \$3.3 billion in revenue a year, yet these were "virtually frozen out of the clean-up and rebuilding" process. Only 1.5 percent of FEMA contracts, worth \$1.6 billion, went to minority businesses, less than the five percent required by federal law.[60] The Army Corps of Engineers similarly awarded only 16 percent of its \$637 million in contracts to minority-owned firms.[61] The first post-Katrina act passed by the state legislature in Mississippi was to eliminate the prohibition of on-land gaming so that offshore casinos could move from barges inland – hardly an act helping near-term recovery – and the state also requested and received a \$600 million waiver to divert funds earmarked for low-income housing to expand the Port of Gulfport.[62] Local parish efforts to distribute food and emergency meals featuring local seafood, pasta, and even wine at a total cost of $13 per meal were rejected by FEMA, which instead issued peanut butter sandwiches and canned beans shipped from corporations in Florida and Ohio for $14 per meal.[63] In some cases, tradeoffs were literally zero sum. The George H. Bush Administration successfully lobbied against a congressional attempt to increase Medicaid coverage for storm victims because the measure would have been funded out of federal oil and gas lease income.[64] Other federal funds went to protect shipping infrastructure and commerce rather than the rehabilitation of the levees.[65]

In fairness, most of the culpability for the disaster's toll probably lies with local and state officials. But the subpar federal aid only exacerbated matters. Far from being anecdotal incidents, analyst Rita J. King went so far as to label Katrina enclosures "The New Big Easy" for corporations, with massive appropriations slated for rebuilding, including $85 billion from Congress, $36.5 billion from FEMA, another $100 million from foreign governments up for grabs.[66] King documents how many of these funds were awarded not based on merit, but due to political connections. For instance, major contracts went to AshBritt, a firm in Florida with close ties to the Bush presidential family; Halliburton, with close ties to Vice President Dick Cheney; Americold Logistics, a lobbyist for the Republican party; and perhaps most interesting of all, a $236 million contract for Carnival Cruise Lines, who won a no-bid contest to provide accommodation for hurricane victims despite the country of Greece offering free use of its ships. As King writes, "the lack of a competitive bidding system in the earliest days and continuous chaos on the Gulf Coast hindered the effort to impose any meaningful accountability on those companies staking claim to the billions in federal recovery dollars."[67] Congresswoman Barbara Lee was even more scathing in her remarks, suggesting that "the aftermath of Hurricane Katrina demonstrated the tragic consequences of having an administration where cronyism trumps competence."[68]

Third, private property developers were quick to procure formerly public land – in some situations, initial displacement was literally forced at gunpoint. When relief and rescue squads were unable to reach people in a timely manner, storm victims were often forced to evacuate under the threat of lethal force by the military, National Guard, or police. Evacuees were literally loaded into vans, buses, and airplanes with no idea of where they were going. In the end, they created a "Katrina Diaspora" that spread to all fifty states.[69] With an estimated 1.5 million people eventually leaving the region, a number of homes owned or occupied by evacuees were declared a nuisance and marked for demolition.[70]

Some actors, however, used Katrina as a mechanism to evict tenants from desirable property, or to raise rent. About 70 to 85 percent of the housing supply of New Orleans received some type of damage, which was "devastating to a city that already had a shortage of affordable housing."[71] Low-income and working-class families that stayed in the area were hardest hit, with rent increases of about 35 percent on average in the first year after Katrina.[72] They were also more frequently evicted from undamaged units, or faced rent hikes for failure to pay rent while evacuated.

Other actors saw the evacuation and rush of available property as an opportunity to rebuild the city according to their own agenda.[73] The Housing Authority of New Orleans (HANO) campaigned to demolish public housing sectors and replace them with a mixed market lot of privately owned homes, commercial enterprises, and a few public housing units – noting that such property "become[s] a prime location for retail and residential development."[74] Part of this new planned development included the building of a Home Depot and other big-box stores. One investigative report from the *Times-Picayune* noted that HANO was allowed to share profits with the stores if their sales exceeded $16 million.[75] Similar trends occurred in Mississippi[76] and Alabama.[77]

A fourth type of enclosure related to schools. Before the hurricane, schooling in New Orleans was primarily public. However, the storm damaged or destroyed more than 100 of the New Orleans Public School (NOPS) district's 128 buildings. At the time, NOPS served roughly 65,000 students; only two percent went to charter schools. After the Hurricane, charter schools acquired rent-free buildings, strategically purchased public school buildings deemed a nuisance, and took advantage of falling property prices. They also aggressively recruited students, and became more selective about other applicants. The strategy worked, and of the 33,200 students enrolled after Katrina, 53 percent were in Charter schools.[78]

These collective enclosures threatened not only to reduce the amount of publicly available housing and schools; they erased a large part of the region's legacy. As legal scholar Alecia P. Long lamented:

> Developers maintain that they are paying homage to the city's history, but they are actually distorting and manipulating it to achieve goals that will repeat past injustices. In the process, their plans may also wipe out the very streetscapes and demolish the small number of remaining structures that played a definitive role in the creation of

> jazz ... We are seeing business as usual when it comes to pushing people and historical structures out of the way in the name of progress. If developers continue to dominate the city's reconstruction efforts – and that looks likely at the moment – the history of [New Orleans'] neighborhoods will be repeated like a long, sad, familiar funeral dirge. Longstanding legacies of racism and displacing the poor will continue, while an upbeat band of developers leads the parade, with their cronies in the government and the corporate media providing a happy, naïve, and self-congratulatory second line.[79]

Indeed, some scholars have argued that the process of reconstruction resembles "Disneyification" rather than historical rehabilitation of damaged areas.[80] Seen in this light, private actors used Katrina recovery efforts to reshape the city around their own narrow goals.

4.2 Exclusion

Katrina recovery perpetuated or created multiple forms of exclusion from: transportation, accessing certain neighborhoods, reconstruction jobs, planning discussions, and insurance relief.

Hundreds of thousands of evacuees were excluded from modern forms of transportation during the evacuation. Those without private cars had to wait for buses, or worse, walk through floodwaters. This meant that thousands ended up stranded inside the city and were crammed into Red Cross shelters or, notoriously, trapped inside the Superdome Stadium or Convention Center. Part of the explanation for exclusionary transportation was that New Orleans Rapid Transit Authority (NORTA) plans were disrupted by the storm, with 197 NORTA buses submerged and ruined along with 24 of its 36 lift vans.[81] Worsening matters, most of the city's 500 school and transit buses were without drivers, who had evacuated earlier in their cars. In the end, only about 100,000 to 134,000 out of 400,000 people were successfully evacuated before the storm. As one commentator put it, "Lacking cars, credit cards, or the several hundred dollars in ready cash necessary to stage any kind of evacuation from the city, many of [the evacuees] stayed behind because they had no other choice."[82]

Evacuees were also physically excluded from wealthier parts of New Orleans. On September 1, 2005, three days after Katrina made landfall, thousands of evacuees reportedly fleeing the Convention Center marched toward a bridge that would take them to safety, only to be turned back by police and sheriffs from Gretna, a predominately white, middle-class neighborhood. As one evacuee told reporters:

> We questioned why we couldn't cross the bridge anyway, especially as there was little traffic on the 6-lane highway. They responded that the West Bank was not going to become New Orleans and there would be no Superdome in their city. The police chief explained that his town feared for its safety from a tide of evacuces.[83]

When a few of the evacuees became desperate and tried to cross forcibly, the police fired warning shots into the air, turning them back.

Another form of exclusion relates to contracts and jobs for reconstruction. Oddly, here, builders took advantage of undocumented (illegal) workers willing to accept extremely low wages, sidelining those seeking competitive wages. After Katrina, President Bush and the federal government suspended the Davis-Bacon Act, passed in 1931 and better known as the Prevailing Wage Law, which set a minimum pay scale for workers on federal contracts.[84] The president also relaxed federal rules concerning the hiring of personnel with proper documents (e.g., waiving Department of Homeland Security sanctions normally imposed on employers who hire illegal workers in violation of the Immigration Reform and Control Act).[85] This led to an influx of low-wage, illegal workers paid below minimum wage rates. As law professor D. Aaron Lacy notes, "The government effectively invited undocumented workers to work for slave wages and offered no incentive for victims to return to these jobs."[86] This hurt both documented (legal) and undocumented (illegal) employees. One employment agency sent African American laborers to construction sites for rebuilding efforts only to report that their laborers were sent home after undocumented workers arrived.[87] Undocumented workers, who will often gather "to where work is abundant,"[88] frequently had to work in hazardous environments without adequate training or protective gear. One study by Tulane University and the University of California at Berkeley revealed that about one-third of undocumented workers used unsafe materials, many slept in tents or moldy hotel rooms and worked 12 to 14 hour shifts, 28 percent said they had difficulty receiving their pay, and only 38 percent indicated they received medical attention when needed.[89]

Another type of exclusion, from decision-making and voting, occurred on multiple fronts. Volunteers and front-line disaster workers, especially women and minorities, were "excluded from the processes of developing and setting priorities for the recovery" and "became discouraged and fatigued at not being heard."[90] One study confirmed that meaningful participatory mechanisms and proper oversight were lacking in Katrina recovery efforts.[91] Moreover, the social and economic upheaval following Katrina meant that a staggering number of vital documents – birth certificates, government identification cards, driver's licenses, passports, bank statements, social security cards – were destroyed or unavailable.[92] As a result, the voting rights of evacuees and displaced residents in local, state, and national elections were "all but ignored," contravening the Voting Rights Act.[93]

Many minorities were excluded from insurance coverage, or denied claims when they did have coverage. At least $40 to $60 billion in insurance claims arose out of Katrina, mostly accessed by middle-class and non-minority survivors,[94] an amount that pales in comparison to the $80 to $400 billion in actual damage. Most minority or poorer households were simply uninsured, and had no access to insurance. FEMA estimated that, as a whole, fewer than 13 percent of households in Alabama, 15 percent in Mississippi, and 47 percent in Louisiana had flood insurance, with even lower percentages for businesses. Economists Masozera and Bailey drew from US Department of Housing and Urban Development data to determine that "city planning districts that had a high percentage of poverty also had a low percentage of flood insurance coverage."[95] Minorities were also less likely to have homeowner's insurance and health insurance. The situation was further compounded by accusations that insurance companies "routinely redlined" black neighborhoods, meaning that residents could not get coverage even if they wanted to or could afford it.[96] Then, many insurance companies denied claims made under valid homeowner's insurance policies, arguing that the losses were either caused by Acts of God (Force Majeure) or were caused by flooding from the levee breach and not the Hurricane itself.[97]

4.3 Encroachment

The encroachment effects of Hurricane Katrina recovery involve three distinct threads: a relaxation of environmental standards, environmental perturbations concerning infrastructure repair and dredging, and a concentration of toxic pollution and debris among some communities.

To expedite emergency repairs to levees and, at a later stage, infrastructural rehabilitation work, local, state, and national environmental standards were curtailed following Katrina. As the Natural Resources Defense Council documented, pro-industry lobbyists were quick to undercut longstanding environmental safeguards "in the name of hurricane recovery."[98] For example, federal requirements concerning the disposal of construction and demolition materials were relaxed, so hazardous wastes intermingled with ordinary items being sent to landfills without protective liners.[99] In another case, state officials relaxed open burning bans to enable the burning of debris in open pits without air pollution controls.[100]

Infrastructure repair and rehabilitation efforts following the storm were so comprehensive, and focused on enhancing human safety, that they often damaged the environment through the use of heavy equipment, dispersal of rock and sand, and dredging. Understandably, immediately following the storm, the priority for planners was stopping floods and closing breaches, not carefully protecting habitats or preserving ecological integrity. In early September, breaches along levees needed to be sealed and many first responders used whatever they could get their hands on, ranging from sheet piling, concrete, and sandbags to debris, furniture, and trees. They also set up an elaborate system of temporary pumping stations to remove water, stations powered by diesel generators that emitted hazardous plumes of particulate matter and air pollution. Particularly large breaches, such as a 700-foot (200-meter) crack in the 17th Street Canal, needed repairing by helicopters (such as the one shown in Figure 4.4) dropping sand bags, and industrial construction machinery such as backhoes, bulldozers, and cranes. Energy- and carbon-intensive materials such as steel and concrete were also needed in the recovery process as key inputs for the reconstruction of some 45 bridges, and hundreds of kilometers of rail, and roads,[101] efforts that undoubtedly contributed to the emission of greenhouse gases through their manufacturing and use. Multiple canals and waterways also needed re-dredging to offset the impacts of shoaling, a process that occurs when sand and dirt clogs a shipping channel and makes it unsuitable for navigation.[102]

A final encroachment effect relates to the concentration of pollution, notably from debris and removal of hazardous materials, in certain communities. In addition to its social and economic toll, Katrina's floodwaters "stirred up a toxic soup"[103] and led to considerable environmental pollution. At least six large industrial facilities were fully flooded in New Orleans and another five partly flooded. Floodwaters containing human excrement and harmful *E. coli* spread diseases such as hepatitis-A and salmonella poisoning, and decomposing human and animal bodies contributed to severe bacterial contamination.[104] In Katrina's wake, other environmental difficulties included contaminated drinking water supplies, oil spills from above ground tanks, flooded sewage treatment plants, hazardous building debris, and contaminated sediment and sludge.[105]

Figure 4.4 A US Navy helicopter drops sandbags on a broken levee in New Orleans, September 7, 2005

Source: US Navy.

Cleanup efforts have unfortunately concentrated much of this pollution in particular landfills or alongside communities of color. As one report to the United Nations warned:

> The US Government's responses to the environmental consequences of the storm have contributed to public health threats. Communities of color – a Vietnamese-American neighborhood in New Orleans and the African American community of Mossville, which is located next to Lake Charles, Louisiana – have had to battle against being subjected to the disposal of massive amounts of debris resulting from Hurricanes Katrina and Rita.[106]

Furthermore, sediment left in the wake of floodwaters contained high levels of arsenic in excess of federal standards, yet the Environmental Protection Agency determined that no cleanup was needed in flooded neighborhoods.[107] As a result, Katrina recovery left severely elevated concentrations of arsenic in soils at numerous playgrounds and schools.[108] Table 4.2 shows at least six schools with concentrations deemed hazardous for human health. Sadly, people repatriated to these polluted landscapes relatively quickly because local authorities were under "enormous political pressure to allow rapid repopulation of the toxin-soaked areas."[109]

4.4 Entrenchment

Entrenchment – aggravating the disempowerment of women or minorities and/or worsening asymmetrical concentrations of wealth – is our final factor at play in Katrina. This section focuses on at least three different dimensions to entrenchment – class, race, and gender – before illustrating how entrenchment resulted in unequal levee protection and left many evacuees trapped in a cycle of "permanent temporariness" or suffering from "Katrina fatigue."

The aggravation of classism and inequality in wealth tops the list of Katrina recovery ills. As Reed asserts:

> [C]lass – as income, wealth, and access to material resources, including a safety net of social connections – was certainly a better predictor than race of who evacuated [New Orleans] before the hurricane, who was able to survive the storm itself, who was warehoused in the Superdome or convention center or stuck without food and water on the parched overpasses, who is marooned in Houston or elsewhere, and whose interests will be factored into the reconstruction of the city, who will be able to return.[110]

Table 4.2 Arsenic concentrations at sample schools and playgrounds post-Katrina Recovery

Sample Location	Street Address	Arsenic Concentration (mg/kg)
Einstein Charter	5100 Cannes	0.4
Mary Bethune Accelerated School	4040 Eagle St.	0.4
Moton Elem	3000 Abundance St.	0.4
Dr. MLK Jr.	2503 Willow St.	0.5
Lake Forest Elementary	12000 Hayne Blvd.	0.5
Lusher Elementary/Middle School	7315 Willow St.	0.5
McDonogh 28	401 Nashville Ave.	0.5
Laurel Elementary	820 Jackson Ave.	0.5
Reed Elementary	2521 Marais St.	0.6
International School of LA	1400 Camp St. (Andrew Jackson Bldg)	0.6
P.A. Capdau Middle School	3821 Franklin Ave.	1.1
S.J. Green Middle School	2319 Valence St.	1.3
Lafayette Academy	2727 S. Carrollton Ave.	10.6
Medard H. Nelson Elementary School	1111 Milan St. (McDonogh 7 Bldg.)	12.4
McMain Magnet Secondary School	5712 South Claiborne Ave.	12.6
Craig Elementary	1423 St. Philip St.	16.1
Drew Elementary	St. Claude Ave. & Pauline St.	20.3
Dibert	4217 Orleans Ave.	22.8
McDonogh Elementary (#42)	1651 North Tonti St.	34.4

Source: Wright, B. (2007). *Testimony before the Subcommittee on Superfund and Environmental Health of the Senate Environment and Public Works Committee regarding Environmental Justice*, July 25. Washington, DC: US Government. *Note*: Shaded areas show concentrations at schools in excess of US EPA standards.

Such sentiments about class have been echoed by studies looking at the close connection between aid money and the income of recipients, showing that homeowners and the middle and upper class were far more likely to receive either insurance reimbursements or government aid.[111] In Mississippi, for example, by 2008 the state had spent almost all (99 percent) of its $1.4 billion in grants designated for homeowners with insurance but only half (52 percent) of its Disaster Recovery Grants for public housing.[112] Statistical analysis of the demographic variables of the displaced has suggested that the ownership of housing, an issue of wealth, might be the single most important factor predicting permanent displacement, more salient than gender, income, or even race.[113] As another group of economists put it, "pre-existing socio-economic conditions play[ed] a significant role in the ability for particular economic classes to respond immediately to the disaster and to cope with the aftermath of Hurricane Katrina."[114]

That is not to deny that Katrina recovery also worsened racial inequality. Racial bias has, for instance, been confirmed in:

- Impacts of the storm and property flooded, with African Americans and other minorities occupying more low-lying land and/or land poorly protected by infrastructure;[115]
- Access to transportation and the ability to evacuate (roughly one in ten African Americans were initially evacuated compared to almost nine out of ten whites);[116]
- Persons forcibly displaced and/or forced into living in temporary shelters (African Americans sometimes lived in cardboard boxes labeled "Katrina ghettos" while white families lived in hotel rooms or on cruise ships);[117]
- Effectiveness in finding new housing – in the disaster area, and across the country – after the storm (white families were reportedly favored two to one by housing companies);[118]
- Ability to return to New Orleans and the Gulf[119] (47 percent of African Americans were unable to return to New Orleans compared to only 19 percent of white households);[120] and
- Financial support and job security in the months following the Hurricane (minorities were four times more likely than white counterparts to lose their jobs after the storm).[121]

As Bullard and Wright explain, Katrina shows that racial minorities:

> Are more likely than all other groups to be underprepared and underserved, and to be living in unsafe, substandard housing. The impact is also cumulative. After a disaster, minorities... suffer a much slower recovery because of the lethargic response by agencies whose participation is critical to their recovery. They often receive less information, are rejected more often for necessary loans, receive less government relief, and endure discrimination and rejection in their search for housing.[122]

Public affairs scholar Camilla Stivers writes that "racism may have shaped policy and bureaucratic decision making and magnified the death, destruction, and misery the storm produced."[123] Sociologist Darnell M. Hunt has gone so far as to argue that "You'd have to go back to slavery, or the burning of Black towns, to find a comparable event that has affected Black people this way."[124]

A third dimension to entrenchment relates to sexism and gender. The Institute for Women's Policy Research has documented that most of the families displaced were headed by single mothers; that women made up the majority of minimum wage workers pushed out by the undocumented workers described above; and that women were generally excluded from multiple phases of the rebuilding process, especially those relating to engineering or physical labor.[125] There were also multiple reported incidents of female rescue workers being sexually assaulted and raped, but virtually no cases of men facing the same threats.[126] Conversely, women were most often the ones doing volunteer and unpaid work. Joni Seager argues that if we had paid attention during Katrina we would have seen "women with children, women on their own, elderly women in wheelchairs, women everywhere."[127] As one scholar noted, "women were primarily responsible for care-work activities."[128] Another remarked that "mopping up is, and always has been, women's work, so it comes as no surprise that large numbers of local women were active in post-Katrina recovery efforts in New Orleans."[129] As a result, US scholars now argue for better integration of gender into emergency management and community practices following disasters.[130]

Taken together, Katrina's relationship with entrenchment was intricate but forceful. One major impact was unequal levee protection *after* Katrina. Congress allocated billions of dollars to the Army Corps of Engineers to fix, upgrade, and rehabilitate about 220 miles of levees and floodwalls, flood gates, pump stations, and canals. However, as of 2011, they never completed the work, spending only $4 billion out of $14 billion.[131] Interestingly, whether by accident or design, it is mostly African American parts of the city of New Orleans still likely to be flooded, because repairs were never implemented. As one study cautions, "increased level protection maps closely correspond with the race of neighborhoods, [with] black neighborhoods such as the Ninth Ward, Gentilly, and New Orleans East receiving little, if any, increased flood protection."[132] As Table 4.3 indicates, many areas of New Orleans either saw no increase in protection after Katrina repair efforts or minimal increases of 0.5 to 1 foot. The community that received the most protection, Lakeview, was mostly white and affluent.

A second impact was that Katrina recovery led to a "continuing disaster" for evacuees, that is still ongoing today. As one social worker managing a Katrina recovery program noted, "In Katrina, what is especially challenging is the disaster started in one place and it's ending in

Table 4.3 Assessment of risk and reliability improvements to New Orleans post-Katrina, 2007

	Average Depth of Flood Water Decrease	Fatalities Decreased	Property Loss Decreased
Lake View	5.5 ft	70%	32%
Upper Ninth	0.5 ft	31%	11%
Lower Ninth	2 ft	29%	4%
Gentilly	0.5 ft	19%	5%
N.O. East (West Lake Forest)	No change	No change	No change
Michoud	No change	No change	No change
New Orleans East	1 ft	83%	24%

Source: Modified from Army Corps of Engineers Interagency Performance Evaluations Task Force (IPET). (2007). *Risk & Reliability Report*, June 20. Washington, DC: Army Corps of Engineers.

another place."[133] This permanent, continual nature of entrenchment surrounding the disaster – the displaced becoming subject to lives of "permanent temporariness"[134] or suffering from "Katrina fatigue"[135] – makes it especially challenging to remedy. Katrina was both a human tragedy and a natural disaster, or as commentators have surmised, an "unnatural disaster"[136] or "catastrophe in the making."[137]

What is particularly disturbing is the tendency for some to naturalize these inequalities as if they were fated by God. As New Orleans Mayor Ray Nagin, an African American, said one year after Katrina, "This city will be a majority African-American city. It's the way God wants it to be."[138] Representative Richard Baker, Republican Congressman from Baton Rouge, was even pithier in his assessment: "We finally cleaned up public housing in New Orleans. We couldn't do it. But God did."[139]

5 Uncomfortable conclusions and implications

Drawing upon these observations, we present five broader conclusions for climate adaptation analysts, researchers, and practitioners at large.

First, political economy processes played in concert across macro scales involving the US Congress, FEMA and the US Army Corps of Engineers, meso scales involving city and state reconstruction planners, and micro scales involving businesses and landlords, as Table 4.4 summarizes. In some instances, factors combined to create a dangerous feedback loop. For instance, the lack of equal levee protection post-Katrina makes it even harder for exposed populations to apply for and afford flood

Table 4.4 The multi-scalar nature of the political economy of Katrina disaster recovery

Scale	Political Economy Predominately Affects	Explanation
Macro: Federal actors (FEMA, US Army Corps of Engineers, US Congress, George W. Bush Administration)	Disbursement of federal funds, relaxation of competitive bidding and wage laws, national insurance coverage, protective infrastructure (levees)	Resources get directed primarily to help large corporations (casinos, cruise line ships, port operators), and homeowners, levy rehabilitation centers on protecting wealthier geographic parts of New Orleans
Meso: States and cities (City of New Orleans, state legislatures, National Guard and police)	Evacuee access to transportation and "safe" parts of the city, property development and rebuilding schemes, prioritization of state funds for ports and canals (and against low-income housing)	Poor, minority, female evacuees less likely to have access to transport, more likely to face forcible resettlement or remain in temporary shelters, private developers use the disaster as an opportunity to promote their own agendas
Micro: Local businesses, neighborhoods and households	Employment opportunities, family assets, social networks and Katrina fatigue	Businesses take advantage of low-wage undocumented workers in reconstruction work, minority evacuees more likely to face rent hikes, eviction, identity theft, loss of their jobs, permanent displacement from the region, and separation from other family members

Source: Authors.

insurance, which in turn translates into less political support (from insurance companies) for protection of their assets. This demonstrates the interconnected, mutually supportive nature of enclosure, exclusion, encroachment, and entrenchment.

Second, Katrina reminds us that in many ways the United States is not as fully democratic, accountable, classless, or meritocratic as it may seem to some. Major, deep-rooted institutional and structural deficiencies related to race, class, and gender exist, and these can become especially pronounced in the midst of disaster. In many ways, Katrina mimicked the sinking of the *Titanic*, where white, wealthy travelers in first class cabins were more likely to survive than those booked into cheaper decks.[140] As McFarlane writes, "That such wealth and poverty co-exist within a First World country illustrated stark contradictions between race, class, markets, government, and principles of fairness and justice."[141] Lowe and Shaw conclude that "just as Hurricane Katrina exposed the depth of poverty and racial inequality in New Orleans, the catastrophe also laid bare the arrested human development in Alabama and Mississippi."[142] One journalist even commented that the images of the failed response made him question America's capacity for world leadership, in part because the country still had such strong divisions.[143]

Third, in the particular case of Katrina, it might be a mistake to look at entrenchment through the isolated lens of class, race, or gender. As Weber concludes, "despite hopes that the influx of Katrina recovery money into the Gulf Coast states would improve the status of the disadvantaged and ameliorate inequalities, the emerging consensus is that the storm and the response to it simply exacerbated these preexisting social ills."[144] In other words, the entrenchment from Katrina was intersectional – cutting across class, race, and gendered dimensions simultaneously. Survivors of Katrina, often poor, female, *and* African American, were frequently confined to future lives of highly irregular employment, homelessness or virtual homelessness, psychological distress, and poor diet and hygiene. As one report noted simply: "Many New Orleans residents faced multiple layers of vulnerabilities based on age, race, or gender, as well as income."[145] Such vulnerabilities are not usually addressed by poverty or disaster programs, yet they keep Katrina survivors trapped "in the basement of extreme poverty."[146] Unfortunately, in the years following Katrina, both media attention and available financial relief have plummeted.

Fourth, Katrina uncomfortably suggests that a negative, underlying political economy can greatly exacerbate the impacts of a natural disaster – perhaps more than a storm itself. Centuries of uneven economic development, traced all the way back to French and British colonization and slavery, and more recently industrialization, political realignment, the privatization or selling off of government assets, reduced social spending, and an increased emphasis on rewarding profit accumulation, left this area of the South extremely vulnerable to a storm.[147] This vulnerability was heightened by a toxic crucible of environmental pollution resulting from the destruction of environmental buffers associated with oil and gas production. Kates and his colleagues suggest that:

> Because disasters tend to accelerate existing economic, social, and political trends, the trajectory for full recovery (preexisting

> population, economy, and infrastructure) is not promising. The large losses in population and employment after Katrina are an accelerated continuation of its 45-year-long decline in population and economic growth.[148]

Even though Katrina ended up flooding some 80 percent of New Orleans, and resulted in as much as $400 billion in damages, the storm was only a Category 3 hurricane when it made landfall, and even then, it was not a "direct hit."[149] Flooding was largely the result of breached and destroyed floodwalls and levees that had not been properly maintained, and bungled bureaucratic delays only worsened what might have been manageable – and certainly predictable – impacts. Katrina reminds us that the "goal of disaster resilient communities cannot be reached until basic issues of inequality and social justice are addressed."[150] As Bullard and Wright warn, "Quite often, the scale of a disaster's impact, as in the case of Hurricane Katrina, has more to do with the political economy of the country, region, and state than with the hurricane's category strength."[151]

Thus, Katrina was a "catastrophe in the making"[152] or a "disaster by design,"[153] but by no means the result of biophysical or natural phenomena alone. Part of this conclusion is positive, for it means that the impact of a disaster can be corrected and perhaps even prevented. Part is negative, however, since the corporate–state alliance operating behind Katrina has since begun to operate in other disasters across the country, becoming a normalized part of disaster response and emergency relief. If planners continue to promote and implement relief efforts the way they did with Katrina, more disasters rather than less will populate America's future. Our conclusion also suggests that even the most admirable or ignoble goals of planners and emergency responders can be distorted by political economy factors.

Fifth, and last, is what Katrina tells us about broader trends relating to markets, politics, and exceptionalism. One lesson here is that market-based or neo-liberal, privatized solutions might not work well in the context of disaster recovery, given that they are amoral (at best) or corrupt (at worst) when it comes to redistribution. Another is that some actors will always see opportunity in crisis, and will use natural disasters as a lever through which they can consolidate their wealth – whether through corruption, the suspension of protections, or strategic gaming of laws and regulations. A final lesson is that it would be a mistake to presume a Katrina-like response is limited to the United States. It can occur wherever there is a nexus of discrimination, poverty, greed, neglect, and alienation – that is, practically anywhere. If Katrina can happen in one of the wealthiest countries in the world – where millions are forced to evacuate a major urban area and billions are spent to recover from a well-predicted storm – it can happen wherever structural inequality, poor planning, and human fallibility coincide.

5

The Perils of Climate Diplomacy: The Political Economy of the UNFCCC

1 Introduction

For all intents and purposes, climate change has resulted in a massive, accidental experiment involving the entire planet and requiring concerted action around the world. In February 2015, scientists at the National Oceanic and Atmospheric Administration's Mauna Loa Observatory recorded an atmospheric concentration of carbon dioxide at 399 parts per million,[1] the highest it has been in hundreds of millions of years. Concentrations of carbon dioxide in the atmosphere could exceed 1,000 parts per million by volume by the year 2050 if trends continue.[2]

Some argue that these elevated atmospheric concentration levels put us in a dire predicament as humans have not been biologically well-evolved to handle crises that do not fit well with our "fight or flight" mentality. Climate change is more like the method of torture known as "slow slicing," or death by a thousand cuts.[3] On the other hand, what distinguishes human survival through evolution is the ability to form social alliances organized around problem-solving tasks.[4] It is yet uncertain if we can organize successfully on a global scale.

As a litmus test of this ability to tackle a dire global problem, the UNFCCC resides at the center of an internationally concerted effort at addressing both mitigation (curbing emissions) and adaptation (coping with impacts).[5] In theory, all countries and actors should be given equal footing in this effort. As the Convention states: "The UNFCCC provides the basis for concerted international action to mitigate climate change and to adapt to its impacts. Its provisions are far-sighted, innovative and firmly embedded in the concept of sustainable development."[6]

With 196 Parties (195 countries and the European Union as a regional economic integration organization), the Convention has attracted almost total global membership since it entered into force in 1994. The Parties have committed to seriously implementing adaptation in tandem with mitigation and improving the state of climate science. They are charged with launching national strategies for adapting to expected impacts, entrusted with providing "financial and technological support to developing countries, and...cooperat[ing] in preparing for adaptation to the impacts of climate change."[7] In fact, the Convention even stipulates that its parties not only consider climate change "in their relevant social, economic and environmental policies and actions," it also states that they should attempt to "minimiz[e] adverse effects on the economy, on public health and on the quality of the environment, of projects or measures undertaken by them to mitigate or adapt to climate change."[8]

In this chapter, we scrutinize the extent to which the UNFCCC meets its obligations to utilize climate diplomacy in a manner that is fair, equitable, and beneficial to all. We find, in short, that its noble intentions still have a way to go. The chapter begins by summarizing a few overarching political economy conflicts in international environmental cooperation. This is followed by a brief history of the struggle to move adaptation up on the UNFCCC priority list. We then provide some examples of how enclosure, exclusion, encroachment, and entrenchment play out in the UNFCCC's mechanisms. The chapter ends with some implications for future climate agreements.

Before readers get the wrong impression, we want to make it clear that neither of us feel that the UNFCCC should be abolished, or that adaptation measures being implemented need to be halted. We support the UNFCCC in principle and practice, to the extent that both of us have participated in its various working groups, attended many of its negotiations (fourteen in total divided across both of the authors), and worked closely with, or advised, its secretariat and many participating parties and observer organizations. Our intent with this chapter is to show that even within this veritable organization, which we admire, a case can be made that the processes of enclosure, exclusion, encroachment, and entrenchment operate. If they can operate here, our thinking goes, they can operate practically anywhere, exhibiting an urgent need for countering them given the scale and scope of climate change itself.

2 Intriguing tensions in environmental cooperation

To better understand how the UNFCCC is influenced by global political economy processes, it is useful to begin with some history. The differences between the nations somewhat sweepingly called the Global North (high income nations like those comprising the European Union, Japan, or the United States) and South (middle- and low-income countries such as Brazil, China, or Mozambique) have been a persistent theme in international environmental negotiations since at least the preparations for the United Nations (UN) Conference on the Human Environment in Stockholm 1972. These differences are rooted in the deep-seated economic inequality that has characterized the post-colonial global economy. The South, as represented by the Group of 77 (G-77), has insisted on linking the global environmental *problematique* to the broader issue of international economic inequality.[9]

This political connection between global problem solving and economic inequality is supported by international political economic concepts such as world system and dependency theories. These theories argue that diplomacy, alliances, and conflict are derivative of the global distribution of wealth.[10] The Stockholm Conference was dominated by developing countries' pursuit of fairer terms of international economic exchange, which eventually became the United Nations General Assembly decision on a New International Economic Order (NIEO) in 1974.

This was the high point of dependency theory in international politics. To avoid suspicions of an international environmental agreement being used to maintain an unequal exchange of cheap labor and raw materials for multinational firms headquartered in developed countries, experts and policymakers in the South demanded state-based controls of capitalist forces alongside reforms of trade and investment rules. Developing countries also called for more foreign aid but with less conditionality.[11]

These political economy related discussions continued at the First World Climate Conference in 1979, organized by the World Meteorological Organization (WMO), the United Nations Environment Programme (UNEP), and the International Council of Scientific Unions (ICSU). The conference declaration pointed to the need for political action to counter climate change consequences for social and economic development and argued that "[s]ince changes in climate may prove to be beneficial in some parts of the world and adverse in others, significant social and technological readjustments may be required".[12] In 1982, UNEP delivered a dismal report on progress following the Stockholm Conference. UNEP concluded that the situation was gloomy in virtually all issue areas. The ten years following the Stockholm Conference had seen "worsening of environmental problems in developing countries arising from the present international economic order which has slowed down their development and the protection of the environment."[13] It was clear that the goals of the Stockholm Conference had not been realized; the international community needed to improve their efforts.

In 1987, a seminal report, *Our Common Future,* paved the way for a world summit covering virtually all of the international environmental concerns of the time. Climate change was not a major issue, but was still ranked as one of the more serious threats for global sustainable development. The report concluded that if a convention on climate pollutants could not be implemented rapidly "governments should develop contingency strategies and plans for adaptation to climatic change."[14] It recognized the vulnerability of groups in developing countries, in particular indigenous peoples and farmers, but focused on reconciling the need to protect the environment with the goal of economic development. The report concluded that environmental problems had social and political dimensions.

In 1989, a UN General Assembly resolution reviewed a large number of additional meetings, which ultimately underpinned the formation of the UNFCCC.[15] Several of these meetings raised concerns about the effects of climate change on poorer countries.[16] Other declarations, referred to in the General Assembly resolution, did not explicitly mention climate but underscored the interdependent relationship between environment and development.[17] As the Intergovernmental Negotiating Committee responsible for the preparatory work for the Convention stressed:

> The extent to which developing country Parties will effectively implement their commitments under the Convention will depend on the effective implementation by developed country Parties of their commitments under the Convention related to financial resources and transfer of technology and will take fully into account that economic and social development and poverty eradication are the first and overriding priorities of the developing country Parties.[18]

Clearly, dependence on the global North for resources and technology was a salient concern.

Another foundational event in establishing the UNFCCC was the Caracas Declaration adopted at the special ministerial meeting of the G-77. It concluded that "poverty and environmental degradation are closely interrelated and that environmental protection must therefore be viewed as an integral part of the development process and cannot be considered in isolation from it."[19] The 1989 UN resolution also referred to other declarations from developing country meetings, which stressed the serious consequences of sea-level rise for low-lying coastal and island states.[20] Contrary to what is often argued, the dominant focus on mitigation was not consistent with the original intentions of the Convention.[21] Since adaptation is primarily framed as a local problem, it has been seen as a mismatch to an international institution aimed at providing "global public goods" and solving "collective action dilemmas."[22] However, in a UN resolution passed the same day as the decision to launch negotiations on a climate convention, the General Assembly urged the international community to provide effective and timely support to vulnerable countries so that they could adapt to these threats.[23]

These varying efforts culminated in the famous 1992 United Nations Conference on Environment and Development held in Rio de Janeiro, Brazil, less formally known as the "Earth Summit." The UNFCCC was one important part of the Rio package. The official rationale for the UNFCCC was that the "largest part of the current emission of pollutants into the environment originates in developed countries, and...therefore...those countries have the main responsibility for combating such pollution."[24]

In discussions about North–South relations on problems such as global warming, the South had for a long time been seen as, and represented itself as, a homogeneous group. This homogeneity was a deliberate political construction sustained by the real vulnerability developing countries felt vis-à-vis their wealthier counterparts in negotiations as well as by the fact that divergences among Northern countries often took center stage at global conferences. The framing of a unitary front by developing countries in the UN climate negotiations can be summarized in five essential features:

1. Global environmental problems necessitate a multilateral response and developing countries will need assistance if they are to be able to participate fully and fairly in the negotiations.
2. Developing countries are especially vulnerable to climate change for a number of reasons, including higher rates of poverty, dependence on agriculture, and geographic location.
3. Industrialized countries have a historic responsibility as they are the predominant source of pollution to date. This, taken together with their greater financial and technological resources, implies that they

should consequently shoulder the burden of the costs of global environmental conservation.
4. Industrialized countries should transfer technology and financial resources to assist developing countries.
5. Any proposals for ameliorating global environmental damage should not jeopardize the right of poorer countries to social and economic development.

These principles form a frame that has united countries from the South in climate change negotiations; they were the basis for the G-77 strategy for UN environmental negotiations. These five core issues are also reflected in the UN General Assembly resolution deciding on a climate convention.[25]

Nonetheless, behind this frame lies a heterogeneity that is equally important for understanding the realpolitik possibilities and constraints for future climate policy as seen from the perspective of the global South. Despite the persistence of the traditional groupings in international negotiations such as the G-77 as one entity, the US and some other OECD (Organisation for Economic Co-operation and Development) countries as another, and the EU (European Union) as a third group, the problem of global warming has created a number of new coalitions that threaten the solidarity and call the framing of traditional groups into question. Tensions are visible in for instance the differences between the interests of the AOSIS (Alliance of Small Island States) and OPEC (Organization of the Petroleum Exporting Countries) as well as the emerging roles of the larger developing states such as India, China, Brazil, and South Africa, whose interests in issues like technology transfer and debt relief have grown in complexity ever since the initial formation of the G-77 group.

The continued existence of the G-77 – as a political entity in the face of these tensions – might be explained by the world system argument that although individual countries might change their relative positions in the system, the system dynamic remains unaltered. And, the G-77's persistence might be explained by a number of other factors. One is a shared understanding among developing countries of global environmental negotiations as a starting point for negotiations over the unequal consumption of resources rather than for the protection of the environment from the excesses of industrial capitalism. The other is that both the weaker and the stronger nations need to coalesce for leverage on issues such as global warming.

Climate change tests these theories, since it is believed that developing countries, on aggregate, are expected to increase their GHG emissions as they pursue continued industrial development. These increasing emissions might provide them with a bargaining chip vis-à-vis developed countries that are pressing for global commitments on mitigating GHG emissions.[26] (As we discuss further in this chapter, the reverse is also true as richer countries have the economic capacity to better cope with the adverse effects of climate change). Climate change will hurt some of the developing countries more than others, either through severe weather phenomena, or since they have fewer resources for adaptation, or both. Thus, developing countries might strengthen their negotiating positions, whereas others have too much at stake to attain a favorable bargaining situation. The emergence of this scenario is dependent on the willingness of the international community to actually take responsibility for the global commons.

3 A crash course in international adaptation diplomacy

Given that cursory history of global environmental politics, this section of the chapter delves into more recent efforts to engender adaptation through international diplomacy. In climate change negotiations, developing countries stress that developed countries have benefitted from the activities that have led to environmental damage. This argument points not only to the link between global environmental change and development in the North, but also to the prospects of future development in the South. This view has even been recognized in Article 4.8 of the United Nations Framework Convention on Climate Change (UN, 2002) which states:

> [T]he Parties shall give full consideration to what actions are necessary under the Convention, including actions related to funding, insurance, and the transfer of technology, to meet the specific needs and concerns of developing country Parties arising from the adverse effects of climate change and/or the impact of the implementation of response measures.[27]

Despite propositions such as these, adaptation has been framed as something positive by nature; it is rarely discussed in terms of winners and losers. This line of reasoning perceives the problem of global warming in biophysical terms and avoids discussion of social and political determinants or consequences. For this reason, one article in the convention dictated that the "Parties shall take full account of the specific needs and special situations of the least developed countries in their actions with regard to funding and transfer of technology."[28]

Interestingly, the Marrakesh Accord, the outcome of the Seventh Conference of the Parties in 2001, included guidance on how to implement the two portal adaptation articles of the Convention. The 2001 decision not only recognized "that the least developed country Parties are among the most vulnerable to the adverse effects of climate change," it also recognized that "widespread poverty limits [the poorest countries'] adaptive capacity."[29]

More pragmatically, the Marrakesh Accord established three funds. First, it created the Special Climate Change Fund, intended to finance projects relating to adaptation via a stated goal of technology transfer and capacity building. The fund also targeted specific sectors for support including energy, transport, industry, agriculture, forestry, and waste management.

Second, the Marrakesh Accord created the Least Developed Countries Fund (LDCF), which would support Least Developed Country Parties (LDCs) in the preparation and implementation of NAPAs, among other initiatives. Both of these funds are financed through fiscal transfers from developed countries and are operated by the Global Environment Facility (GEF). The GEF is a multibillion dollar organization set up in the early 1990s to assist developing countries with the costs of environmental protection. The GEF is administered by the World Bank and monitors progress toward the goals of sustainable development. The GEF is not without its critics or problems, the most obvious of which is that many countries did not honor their pledges to the institution. A second is that developing countries has argued that GEF assistance was biased toward problems that the North defined as pressing (e.g. mitigation of greenhouse gases), and overlooked local environmental problems and priorities.

Third, the Marrakesh Accord established the Adaptation Fund (AF) to finance concrete adaptation projects and programs in developing countries that were Parties to the Kyoto Protocol and are particularly vulnerable to the adverse effects of climate change. The Adaptation Fund was financed with a share of proceeds from the Clean Development Mechanism (CDM) and some other sources of funding. Two percent of the proceeds from certified emission reductions (CERs) issued via the CDM are provided to the Fund. Thus, the fund is not financed primarily with development aid, but through market transactions. It is, therefore, not directly attached to the national budgets of developed countries.

These three funds are intended in part to ensure that the wealthier countries meet their moral obligation to shoulder the costs of adaptation. The same logic is reflected in calls to use the principle of historical responsibility to muster support for burden sharing for adaptation finance. The concept of historical responsibility in UNFCCC negotiations implies that individual countries have an obligation to take action on climate change based on the relative effects of their past emissions.[30]

The concept of responsibility has been used to assign moral liability in two different ways. It has been used to argue for an operationalization of the "proportional allocation" of mitigation and financing commitments based on modeled historic contributions. Therein, heavily industrialized countries' carbon debt provides a basis for adaptation liability.[31] But responsibility is also used "conceptually" to recognize the historically induced unequal distribution of capacity to act and infers a relatively greater responsibility to act among countries who have benefitted from colonialism or economic dependency relationships. Yet, responsibility does not require models that specify every country's burden. Conceptually, historical responsibility has been recognized as a principle in the Convention since the Cancun meeting of 2010.[32]

A few other items of merit deserve mentioning. Since the Bali Action Plan adopted in 2007, Parties now agree that future international climate change cooperation should include adaptation as an important building block. The 2010 Cancun Adaptation Framework introduced a new national adaptation planning process for developing countries and established an Adaptation Committee to provide technical support and information, and also highlighted insurance issues. Despite these gains, deciding how to deal with the distributive aspects of adaptation to climate change remains a hotly contested issue in international negotiations. At the COP 15 in Copenhagen, developed countries pledged to mobilize $100 billion annually from 2020 onwards to support climate action in developing countries. Concurrently, the Green Climate Fund was established in 2010 as the new operating entity of the Convention's Financial Mechanism.[33] The Green Climate Fund is expected to become a key player in the climate finance landscape. At the COP 20, held in Lima in 2014, the fund had exceeded its initial capitalization target of $10 million. Still, controversy erupted at Lima over the scope of the Intended Nationally Determined Contributions to a climate agreement, with developed countries advocating for a sole focus on mitigation, and developing countries seeking the inclusion also of adaptation and the means for implementing climate actions, that is, financing, technology, and capacity building.

Thus, the continued capitalization of climate finance remains a critical issue in the negotiations. As Huq, Roberts, and Fenton concluded, "inadequate and poorly defined rules on climate finance stand as a major obstacle to building trust between the materially powerful and the vulnerable poor countries."[34] Financing climate action is also at the heart of the conflicts surrounding the lead up to a new international climate agreement expected to be concluded in Paris in December 2015.[35]

4 Exposing the political economy of the UNFCCC

It is evident that political economy aspects have influenced climate change diplomacy. To what extent can we find political economy reflected in international policymaking on adaptation? Here, we will provide a few examples of how the processes of enclosure, exclusion, encroachment, and entrenchment occur in the UNFCCC's adaptation work. In sum, the UNFCCC engenders enclosure by eroding the autonomy of public and even some private actors to engage and act on climate change matters. It perpetuates exclusion by marginalizing some stakeholders and groups in the COP process. By (understandably) prioritizing the process of adaptation, UNFCCC-funded projects have been known to encroach upon ecosystems and result in environmental damage. Lastly, the UNFCCC process risks contributing to global inequality – fostering macroeconomic entrenchment – by failing to police funding pledges and by failing to facilitate the transfer of state-of-the-art adaptation technologies.

4.1 Enclosure

The UNFCCC facilitates enclosure by commandeering decision-making roles that formerly belonged to governments. Given national government inaction on the topic – the nation wide actions pledged so far are still dismal in light of future adaptation needs and current emission trajectories. – this type of enclosure is understandable, perhaps even laudable. The element of governments' inadequate action is in a sense a larger source of injustice since, in effect, the result is higher emissions and adaptation burdens. But the UNFCCC response to this vacuum also leads to some pesky political economy problems.

In a way, the UNFCCC represents a relatively new form of governance – an international treaty with an established secretariat and a host of experts and bureaucrats – that is essentially a node of power beyond the state.[36] The UNFCCC is a parallel bureaucracy that has appropriated, or enclosed upon, what used to be the domain of governments. Even with the best of intentions, and some strong justifications for intervention in the face of governmental *lack* of action, an international regime like the UNFCCC requires a framework of rules, expectations, prescriptions, and conduct in order to function.[37]

A paradigmatic example here is the Least Developed Countries Fund (LDCF), the largest global fund for adaptation projects (by volume), managed by the Global Environment Facility. During the Fourteenth Conference of the Parties at Poznan, Poland, in 2008, some least developed countries "expressed their frustration" at this structure, at the slow pace with which projects were allocated funding, and at the "long and complicated" process of requesting money from the fund.[38] Figure 5.1, for example, shows the complexity of a typical LDCF project cycle for the United Nations Development Program, one of ten implementing agencies. The project cycle involves almost a dozen internal "administrators" and requires 14 to 17 *months* to navigate.

Three independent evaluations of this process suggest that it results in friction and lags behind in implementing adaptation projects. The first, conducted by the UNDP in 2009, noted "justifiable dissatisfaction" among participants of the LDCF "concerning the lengthy time periods and complex procedures required to move from the NAPAs to concrete projects. In some cases, these procedures have led to time lapses of several years before projects get off the ground."[39] That review noted, for example, that projects took an average of 471 days to begin due to "bottlenecks" and the "many stakeholders and consultations involved."[40] It found that even the preparatory phase required "a lot of work" that ended up being "demanding" for country offices; the review cautioned that GEF requirements and project criteria were "complicated" and "poorly understood."[41] It lastly noted that the co-financing requirement of the LDCF meant that some countries did not have the resources needed to commence the projects.[42] A second review, conducted by the Danish Ministry of Foreign Affairs in 2010, concluded that "in order for the LDCF to play a complementary role to the emerging other climate change financing mechanisms greater responsiveness and flexibility of procedures will have to be introduced to ensure lack of duplication and complementarity."[43] And a third review, from the non-profit Climate Change Forum in 2010, criticized management structures at the LDCF that were "too complex," accused implementing agencies such as the World Bank of adding "further bureaucracy to the process," and concluded that "rules and structures make accessing funding difficult...and time-consuming."[44]

Within this complex bureaucratic structure, some departments and bodies have extended their domain over particular issue areas. For example, most negotiated vulnerability and adaptation action items have been

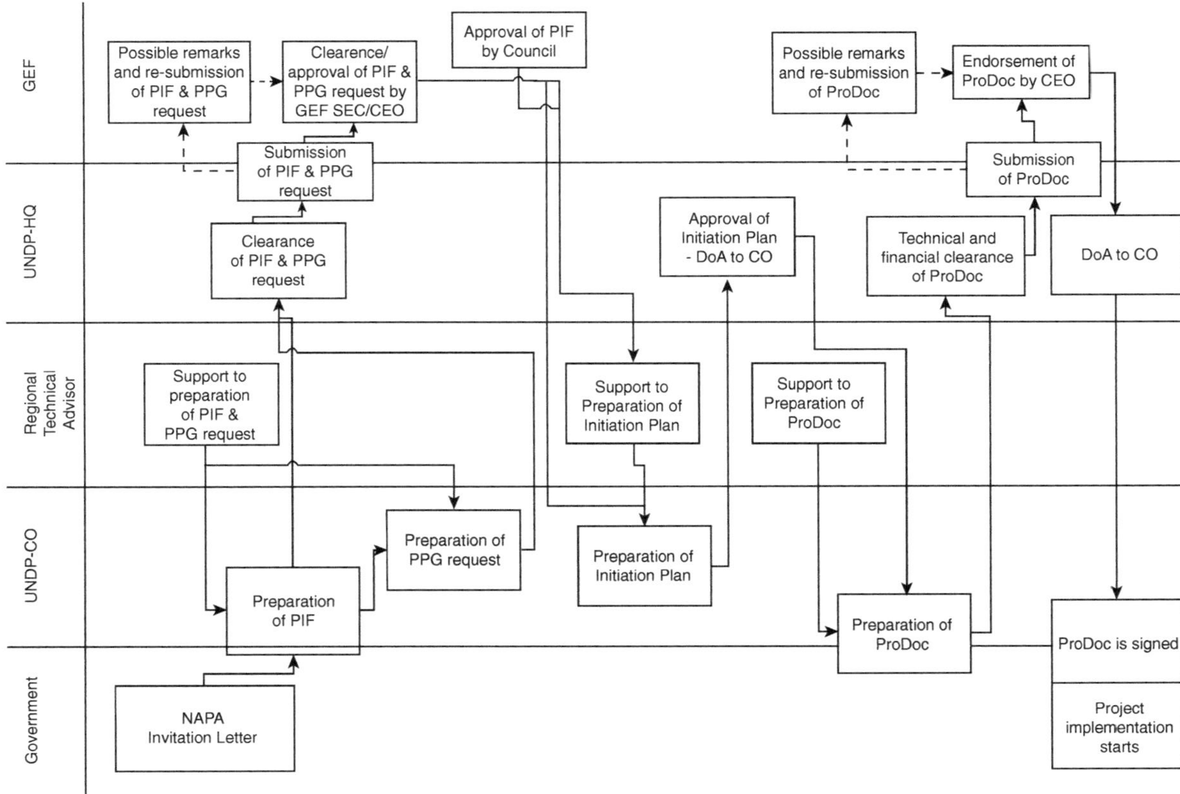

Figure 5.1 Process flowchart for a UNFCCC adaptation project

Source: United Nations Development Programme.

addressed by the Convention's Subsidiary Body for Implementation. It has reached agreements on a number of issues, including: How Parties should fund and support developing countries' assessment of impact, vulnerability, and adaptation needs; how to spur capacity building, training, education, and public awareness; implementation of concrete adaptation activities; promotion of technology transfer to assist adaptation measures; and regional workshops to exchange experiences of adaptation.

Put another way, establishing the scope of international organizations is subject to constant political bargaining as well as bureaucratic and legal disputes between countries with unequal resources. What poorer countries tend to lose in negotiations is autonomy over planning and implementation processes, as well as project finances. When UN organizations assume monitoring roles, they often enclose on the public sectors of poor countries. Thus, climate negotiations involve not only struggles between countries with different interests, but with international organizations and bureaucracies inclined to broaden their mandates and scopes of action.[45]

4.2 Exclusion

Discussions and negotiations within the UNFCCC are rarely representative and often asymmetrical and exclusionary. The result is that only particular types of adaptation projects are implemented and the overall negotiation process tends to benefit some powerful parties at the exclusion of others. For instance, vulnerable developing countries are not equal partners in international negotiations and lack the ability to advance their own interests.[46] Such countries have a small market share of the global economy, are not members of the Organization of Economic Cooperation and Development (meaning they are restricted from joining the International Energy Agency and other groups), and can only afford to send small delegations to key climate meetings.[47]

The result is that industrialized countries such as the United States and industrializing countries such as Brazil, China, and India are able to wield disproportionate influence. At the fourth Conference of Parties (COP) in Buenos Aires, Argentina, in 1998, the delegation from the United States consisted of 83 people. The average delegation from Africa, by contrast, had between two and four people. At the 13th meeting in Bali, the United States had a delegation of 224, whereas 79 countries had less than seven people to cover the vastly expanded negotiation agenda. Villanova Law Professor Ruth Gordon argues that "African countries seemed to have no, or a weak, position and stance and have received few concessions."[48]

In parallel, at the fifteenth COP in Copenhagen, Denmark, in 2009, delegates from the United States and European Union boasted 1,200 limousines and 140 private jets; delegates from emerging economies in Asia and Africa, by contrast, arrived mostly by taxi and flew economy class.[49] This may be disturbing for moral reasons, but it also exemplifies the vast differences in economic power with tangible consequences. As energy and climate scholars Stephen H. Schneider and Janica Lane lament, "the most marginalized groups tend to have little political and economic power and hence have little influence in the decision-making process."[50]

Philosopher Henry Shue argues that disproportionate representation in negotiations is deplorable not only for being unfair, but also because it results in "dirty development" and higher rates of emissions. As he writes:

> Wealth provides bargaining strength because it constitutes resources that can be drawn upon in the absence of any agreement, making the wealthy bargainer willing to pay relatively less to reach agreement. The wealthy party can draw upon her wealth to survive non-agreement. In this case, however, non-agreement means dirty development rather than clean development... the largest of the poor nations will presumably proceed with what they would have done anyway: the most cost-effective available development strategy in which the costs of, for example, emissions of greenhouse gases are ignored, as they were during the development of now wealthy countries.[51]

Wealth enables large delegations, which clearly benefits certain countries' positions in the negotiations. Ironically, it also provides countries with a bargaining chip, as they can better cope with the effects of a non-agreement, at least in the near future. For instance, OPEC, led by Saudi Arabia, linked its demand for compensation for lost revenues if climate policies led to reduced oil sales to Articles 4.8 and 4.10. The OPEC countries argued that compensation from the Annex 1 countries was needed to allow oil producers' single source economies to make investments needed to diversify their activities, or adapt to changing climate economics.[52] To be fair to such wealthier nations, though, the marginal position of adaptation in the Convention and the meager progress in operationalization of its supporting mechanisms was also a result of disunity within the G-77.

The UNFCCC, by acting as a powerbroker, is able to set agendas that usually benefit only some Parties. This is particularly apparent in the use of knowledge production to underpin countries' positions in the negotiations. Indeed, since the UN Conference on Science and Technology for Development in 1979 it has been well understood that the majority of developing countries lack the necessary scientific infrastructure or capacity to support their development aspirations. This has clearly been detrimental for developing countries in negotiations that heavily rely on scientific data, especially concerning the IPCC, which works closely in tandem with the UNFCCC process.[53] As Gordon explains:

> IPCC working groups are almost completely composed of American and European scientists. With a smaller scientific and technological base, southern nations were, and perhaps still are, at a distinct disadvantage in influencing the scientific debate. For unindustrialized nations with few if any scientific institutions, participation in this aspect of the process is non-existent. Thus, at least initially, the climate change discussion was entirely driven by industrialized nations, despite the potentially devastatingly adverse impacts on the South. Moreover, science has been used to implicate southern countries in the climate change quandary, either by maintaining that their future emissions are increasing and would make current action by industrialized nations counterproductive, or by failing to distinguish between southern survival emissions and northern luxury emissions.[54]

In other words, developing countries often lack the scientific or political capacity to advance their own agendas.

The UNFCCC secretariat recognizes the knowledge disparity problem. One instrument to try to address it, even if marginally, is the side events at the negotiations, which have mushroomed in recent years. There are now so many that it would be impossible for a single delegate to attend them all. Figure 5.2, for instance, shows a small selection of the side events available to participants at the COP20 in Lima, Peru. Some developing countries have started justifying their attendance at the COPs on the basis of building capacity and accessing research ideas from these side events.[55] Unfortunately, despite high demand, instead of being prioritized by the Parties to the convention they have in recent years reduced the funding for hosting these side events.

Marybeth Long Martello and Sheila Jasanoff have even argued that the COP process under the UNFCCC is hegemonic since it treats all emissions alike – regardless of their source – so that, in essence, it penalizes subsistence activities just as severely as luxury ones. Farmers producing greenhouse gases from rice paddies are to be held just as accountable as

Figure 5.2 Announcement board for side events at COP 20 in Lima, Peru
Source: Authors.

yacht owners on a luxury fishing trip, a policy that conceals an intention of dominant parties not to distinguish emissions sources.[56]

The UNFCCC also fosters exclusion within nation-states. Even when national and local stakeholders retain power over climate change projects, the UNFCCC process can legitimize centralized decision-making. For instance, the Reducing Emissions from Deforestation and Forest Degradation (REDD+) mechanism – which aims to "compensate developing countries for forest emissions reductions,"[57] making it a mix of mitigation and adaptation – promises substantial remuneration[58] to nations that can better manage their forests.[59] Though the program has given some indigenous groups and forest managers a platform to voice their views,[60] it has also incentivized national governments to focus on short-term gains and practices deemed useful by the international community.[61] Perhaps predictably, local stakeholders throughout the globe have reported that REDD+ pilot projects have violated free, prior and informed consent (FPIC) principles.[62] In Peru, indigenous groups have reported land tenure issues as a result of new, centralized forest management programs prompted by REDD+ funding.[63] Thus, a process intended to elevate indigenous people's rights and respond to climate change has inadvertently excluded their priorities from many projects.

4.3 Encroachment

Encroachment, as part of the UNFCCC, surfaces in adaptation projects, resulting in deforestation or changes in land use. Projects sometimes involve replacing native forests and grasslands with fast growing trees that absorb more carbon, or enhance resilience, but they also divert water from local residents, require pesticides, and displace communities from land.[64]

Though UNFCCC efforts have bolstered adaptive capacity around the world, some projects have required large infrastructural modifications that have all but decimated ecosystems. In Bhutan, for instance, a project draining glacial lakes has proceeded much more slowly than expected, has accelerated erosion, and impinged upon water supply. Thorthormi Lake is so remote that all infrastructure and materials must be flown in by helicopter. Heavy machinery like excavators is needed to move boulders and manage drifting icebergs and strong winds.[65] Figure 5.3 shows volunteers hauling rock across a streambed in 2010.

In the Maldives, a UNFCCC-supported "Safer Island Development Program" has focused on identifying and populating a range of islands that would act as "safe havens" for people forced to migrate before or after natural disasters induced by climate change. Its government sponsors promoted human-built infrastructure such as sea walls and desalination plants to cope with rising sea levels and saltwater intrusion of freshwater supplies. The program has involved widespread disturbances to human

Figure 5.3 Bhutanese volunteers draining Thorthormi Lake in 2010
Source: Authors.

communities and natural ecosystems. Coastal protection measures under the SIDP, such as dredging to create sandbars or erecting seawalls, has unintentionally reduced the flow of nutrients to coral reefs, weakening a natural shield against storm swells and surges. Coastal communities have removed vegetation to make way for the expansion of ports and harbors, and mined sand so that it can be used in construction.[66] Hulhumalé in Malé Atoll, intended to act partly as a home for climate refugees, had to be reclaimed from the sea by converting a natural coral reef and lagoon into a two-square-kilometer city. Sand ridges were repeatedly leveled during land reclamation. The dredging took five years and cost $32 million to complete to say nothing of the expense of building and populating the island, depicted in Figure 5.4. Dhuvaafaru Island in Raa Atoll was also unveiled in March 2009. Formerly an uninhabited forest, the entire 40 square hectare island was raised and a new village built for the 4,000 survivors from Kandholhudhoo, an island destroyed by the 2004 Tsunami, at a cost of more than $40 million.

The encroachment effects from these adaptation projects are not limited to the environmental sphere. As Campbell, among many, points out, "promoting adaptation measures that try to enhance safety nets from forests sounds like a potential win-win, but by virtue of keeping forests rather than converting them to more productive activities may reduce the longer term ability of rural households to get out of poverty."[67] Synergetic effects between policy areas often involve tradeoffs, which

Figure 5.4 Workers erect a housing unit on the reclaimed island of Hulhumalé in the Maldives, June 2010

Source: Authors.

are obscured in win-win rhetoric. For instance, adaptation measures that seek to heighten agricultural development and improve livelihood security might do so only at the cost of reduced land and reduced income for some farmers.

4.4 Entrenchment

Because the UNFCCC operates at the global level, most of its entrenchment is grand in nature and international in scope; it relates to a discernible lack of fair (and promised) funding for adaptation as well as limited technology transfer, which keeps large parts of the South locked – entrenched – in technological inferiority and macroeconomic inequality.

One dimension of entrenchment is financial: funds are not relegated to UNFCCC adaptation projects despite commitments to the contrary. Some actors in the global South have suggested that technological transfer and monetary flows connected with CDM and Joint Implementation follow the line of most foreign investment and go to the larger economies of developing countries, leaving Least Developed Countries on the margin.[68] Since poorer countries do not produce high enough quantities of emissions for effective projects that render good returns in emission credits, they run the risk of being left out of investment calculations.[69] The CER technically belongs to the host country where the CDM investments are made, which is a non-Annex 1 country. Consequently this funding comes from developing countries, and not the Annex 1 countries making investments in CDM. Countries, like Brazil, China, and India, which host the overwhelming share of CDM projects, look at the CDM levy as a South-South transfer, a "solidarity payment" from developing countries to the LDCs.[70] Similarly, Okereke has criticized the Adaption Fund's funding scheme as built on voluntary CDM investments rather than a "polluter pays" principle, a revered principle in developing countries.[71] Sterk, Luhmann, and Mersmann foresaw the need for adaptation in developing countries at roughly \$50 billion by 2020, whereas the World Bank estimates \$75–\$100 billion annually between 2010 and 2050 with a 2° C warming.[72]

However, the likelihood of this appeal becoming reality is dubious given the track record to date. As of 2014, wealthy nations have donated a paltry \$10 billion per year to developing countries via the Green Climate Fund.[73] As of mid-2014, developed countries had only disbursed \$1.3 billion pledged via four other major multilateral adaptation funds (see Table 5.1).

Table 5.1 Funds disbursed to date from UNFCCC adaptation funds (in millions of US dollars)

		2007		2012		Mid-2014	
Fund	**Goal**	**Pledged**	**Disbursed**	**Pledged**	**Disbursed**	**Pledged**	**Disbursed**
Strategic Priority for Adaptation (SPA)	Pilot projects that address local adaptation needs and generate global environmental benefits	50	15	–	50	–	50
Least Developed Countries Fund (LDCF)	Implementation of most urgent adaptation projects in least developed countries, based on National Adaptation Programs of Action (NAPAs)	163	12	537	334	880	726
Special Climate Change Fund (SCCF)	Activities aimed at adaptation as well as three other purposes: technology transfer, economic diversification, and support in key sectors	70	6	240	162	333	299
Adaptation Fund (AF)	Concrete adaptation projects in developing countries that are particularly vulnerable to the adverse effects of climate change	–	–	350	165	350	226
Total		283	32	1,128	711	1,563	1,301

Source: Modified from Sovacool, B.K. (2013). *Energy & Ethics: Justice and the Global Energy Challenge*. New York, NY: Palgrave Macmillan, updated to 2014.

In the absence of ambitious emission reduction accomplishments, the need for adaptation finance will increase. However, private financing is unlikely to cover a substantial proportion of this need.[74] In another overview of this finance landscape, Buchner et al. estimated annual global climate finance flows at $343–$385 billion in 2010/2011.[75] Public funding amounted to $16–$23 billion whereas private actors raised $217–$243 billion in both developed countries ($193 billion) and developing countries ($172 billion). This all looks promising until one realizes that the vast bulk of the finance ($330.7–$369.3 billion) targeted mitigation activities. Only a small proportion targeted adaptation.

Even when money is awarded, it usually comes with hidden strings attached. As one example, because the LDCF is supposed to prioritize "equitable access" for all participating countries, individual projects have a "ceiling" on the amount they support. For instance, from 2001 to 2006 the cap on LDCF projects was $3.5 million, in 2008 it was raised to $6 million, in 2010 it was increased to $8 million, and today it is $20 million. Though the LDCF has a mandate to finance the full additional cost of adaptation – without a requirement for matching co-financing – in practice the ceiling inadvertently requires hosting governments to co-sponsor projects or find other institutions such as the UNDP or FAO to "match" contributions. Moreover, because the LDCF is voluntary, it is only replenished when donor countries decide to be generous, making it difficult to accurately predict the amount of resources available to countries over long time frames.[76]

Worryingly, in terms of *what* they fund, UNFCCC projects usually focus on technology or infrastructure but do not target minorities or the most vulnerable. Remling and Persson analyzed 27 full project proposals from the Adaptation Fund that had been approved by the end of 2012. Some proposals (e.g. Honduras) provided information on the lack of adaptive capacity created by poverty and livelihood situations. Only two proposals, however, reflected on gender and ethnicity aspects as relevant for their selection of adaptation location (Ecuador, Lebanon). Beyond these examples, the authors found "no reference at all to powerless and politically marginalized groups."[77] As they concluded: "None of the proposals addresses sub-national inequities, entitlements, or power structures."[78] Again, as at the national level, "vulnerability" is still mainly framed as an environmental problem, with less reference to its social, cultural, political, or ethical dimensions. Thus, whether UNFCCC projects address the weakest and most vulnerable parties remains disturbingly dubious.

Technology transfer is another channel of entrenchment. The UNFCCC defines it as "a broad set of processes covering the flows of know-how, experience and equipment for mitigating and adapting to climate change amongst different stakeholders such as governments, private sector entities, financial institutions, non-governmental organizations (NGOs) and research/education institutions."[79] Consequently, technological transfer is not only concerned with the transfer of equipment and hardware but also with the soft aspects of technology use, such as knowledge, capacity, skills, and the processes involved in using technologies. This latter definition would include awareness-raising, education, and training. In addition, it would involve organizational issues such as ownership and institutional arrangements within societies. In other words, technology transfer is not only about transferring hardware from industrially advanced countries to less-developed nations. It is about supporting the use of locally existing equipment, infrastructure, and practices.[80]

The topic of technology transfer has become a longstanding and salient one in climate policy literature.[81] Within the UNFCCC itself, transfer of technologies for adaptation came into the fore at the Cancun Agreements in 2010. These Agreements included decisions to create the Climate Adaptation Framework and the Technology Mechanism, which pushed Parties to scale up their efforts at providing technologies for adaptation as well as technical and financial support. Moreover, the UNFCCC Special Climate Change Fund has two separate windows for funding of adaptation and funding of "technology transfer for adaptation," and the negotiations have two separate tracks of planning frameworks for developing countries under the UNFCCC process. In addition to the NAPAs, national Technology Needs Assessments shall "identify and determine the mitigation and adaptation technology priorities of Parties other than developed country Parties."[82]

Despite this attention to the unique needs and capacities of poor countries, technology transfer under the UNFCCC has contributed to entrenchment in at least three ways. First, the bulk of funds have (oddly) gone to industrialized rather than developing countries. The thinking is that poor countries are more likely to accept technology transfer if devices or processes are available cheaply or accessible through the public domain. Middle-income countries have engineering firms and programs for Research, Development, and Demonstration (RD&D), and are therefore seen as the best conduit for adaptation funding.[83]

Second, participating countries can hoard their innovations, limit investments in capacity domestically, lack the training or institutional capacity domestically to appropriate technologies or foreign aid, or restrict investments. Thus, unequal positions in the world economy "entail ... selective transfers of technology and know-how while making sure that the impetus for the direction of technological change" remains with developed countries.[84] These issues definitively hamper the rate of technological development.[85]

Third, technology transfer policies often encourage the governments of indebted countries to create more favorable conditions for foreign investors and transnational corporations. Research on the unequal exchange of foreign investment dependence and the environment have revealed that increased levels of foreign investment in less-developed countries do not necessarily lead to the spread of cleaner technologies and production processes, but instead risk being an instrument for polluting or resource depleting extraction, ultimately threatening human development.[86] Thus, the selectivity of technology transfer cements a pattern of unequal exchange between rich and poor countries.

Indeed, the situation of dependence, entrenchment, and lack of technology transfer mimics another aspect of the UNFCCC (though not dealing with adaptation): the Clean Development Mechanism, or CDM, which involves mitigation and low-carbon energy systems. There, the intellectual property rights for many high-tech low-carbon technologies lie with Western technology firms; countries in the developing world have not been able to guarantee protection of their intellectual property, and technology transfers simply have not occurred. For instance, wariness about weak patent protection has deterred the distribution of innovative technologies from the United States to Asia, as US firms believe they would be at a competitive disadvantage if they shared technology. In addition, host companies in developing countries have shown a reluctance to acquire technology that they believe competitors in other nearby countries could freely copy in their own markets. Such concerns have complicated (and at times prevented) plans to disseminate efficient industrial boilers, fluidized bed combustion, coal gasification systems, and various end-of-pipe pollution abatement technologies such as carbon capture and storage in addition to hydrogen fuel cells, wind turbines, solar panels, and batteries for electric vehicles.[87] The problems with the CDM suggest that elements of entrenchment are systemic within the UNFCCC, and perhaps not isolated to adaptation alone.

The situation with entrenched dependence has convinced many members of the South to abandon the UNFCCC approach to technology transfer and simply transfer systems from South to South. For instance, promisingly, South–South trade in renewable energy products is expanding. South–South trade in the core products of solar, wind, hydropower, and solid biofuels technologies grew at 26.7 percent in the period 2004–2011, compared to only 9.7 percent growth in the global manufactures trade.[88] In South and South East Asia, Bangladesh and India continue to move ahead in the dissemination of distributed renewable energy systems to electrify off-grid areas as well as provide more reliable energy in rural areas. Electrical systems in the form of solar lanterns, solar home lights, and mini/micro grids and improved biomass cook stoves for thermal heat have been attempted in these countries. Today, the national capacity of renewable-energy-based electricity in Bangladesh is approximately 403 MW, out of which almost 158 MW consists mainly of solar PV systems.[89] We expect that similar South–South patterns will appear in adaption cooperation should the present slow pace of technology transfer continue.

5 Uncomfortable conclusions and implications

The UNFCCC process fills an important void in launching climate change policies, capacity building, and knowledge around the world. Many critics have proclaimed its demise in favor of other fora like the Group of Eight consisting of the wealthiest countries, or the Group of 20, including most of the largest economies and accounting for most of global emissions. Such fora might fit the mitigation framing, but would risk even further exacerbating the unequal exchange we highlight with adaptation. With this in mind, we offer four conclusions.

First, this chapter has pointed at some of the existing or potential risks that the UNFCCC can aggravate. In this way, the UNFCCC is complicit with a tendency for global environmental negotiations to mask unequal access to environmental resources or restrict access to technological resources. As Table 5.2 indicates, these processes cut across numerous scales. The UNFCCC participates in enclosure by setting rules and negotiations that benefit some parties but not others. It relies upon a COP process replete with unequal exchanges of knowledge and negotiation power. Its projects encroach at times upon sensitive biodiversity areas, and its patterns of funding and technology transfer can entrench the economic and technological divide between North and South.

Second, concerns related to financing and technology transfer threaten to atrophy the heart of a future climate agreement. Though we concur that a new climate agreement is urgently needed, it must to not only incentivize developing country mitigation commitments but truly seek to muster historical responsibility to build resilience and

Table 5.2 The multi-scalar nature of the political economy of adaptation in the UNFCCC

Scale	Political economy predominantly affects	Explanation
Macro: International negotiations	Developing country delegations	Vulnerable developing countries are not equal partners in international negotiations and have relatively weaker capacity to advance their own interests in processes
Meso: National institutions including legislation and implementing bodies, country populations	Countries' decision-making authority; Human development goals; Economic position in the world economy	Limited and conditioned technology transfer keeps the South locked – entrenched – in technological inferiority and macroeconomic inequality
Micro: Local communities	Farming, forestry, livelihood security, adaptive capacity	Adaptation projects can interfere with healthy ecosystems and keep developing countries mired in technological inferiority

Source: Authors.

human development in vulnerable countries. Adaptation finance needs to be secured, and greater clarity on the terms of finance and technology transfer needs to be reached. In particular, the institutional requirements must place a greater priority on distributing resources to developing countries. Technology transfer needs to consider not only formal requirements and hardware, but also the soft aspects of technology use, such as knowledge, capacity, skills, and processes involved in using technologies as well as awareness-raising, education, and training, in addition to organizational issues such as ownership and institutional arrangements within countries. The UNFCCC must put adaptation, poverty alleviation, and human security on a par with mitigation.

Third, discussions and negotiations within the UNFCCC need to become more symmetrical and inclusive. For instance, vulnerable developing countries are not equal partners in international negotiations and frequently lack the ability to advance their own agendas. At the very least, the UNFCCC can strengthen the processes it has initiated to, on a small-scale, address disparities in negotiating resources and the knowledge-producing capacities of climate meetings. One such example is to strengthen the side events as a site of information sharing, capacity building, and networking.

Fourth – before readers start throwing bricks at the UNFCCC headquarters – the right question to ask might not be "why doesn't the UNFCCC work better?" but "given what we should expect about politics, how does the UNFCCC function so well?" In the face of the current lacunae over global climate policy at the moment, one could justifiably wonder why the whole system does *not* perpetuate the processes of enclosure, exclusion, encroachment, and entrenchment in more severe ways. The international community has at least made an attempt to slow climate change. Further, though its pledges have not been fulfilled, the international community has raised some $1.3 billion for adaptation projects around the world – no small sum even if it is wholly insufficient. And, the UNFCCC has raised these funds relatively quickly – in about a decade – and on a scale involving almost four-dozen countries and hundreds of partnering institutions. Given that it relies on goodwill, it is perhaps amazing that the UNFCCC works as well as it does.

6
Principles and Best Practices for Climate Change Adaptation

1 Introduction

One of the central dilemmas in politics, public policy, and economics is that markets "work" only at distributing certain types of goods. They tend to be efficient at distributing private goods such as bicycles or hamburgers – where property rights can be completely defined and protected, where owners can exclude others from access, and where property rights can be transferred or sold[1] – but less effective at common pool resources such as fish in the high seas or grasslands for grazing, which require agreed-upon rules or sanctions. Unfettered economic markets are almost always completely ineffective at distributing these common pool resources. Designing workable, viable management of common pool resources is "tremendously difficult," since in many cases success depends upon creating an "inverse commons" where material scarcity is not a concern and each additional user increases value rather than diminishes it.[2]

Luckily, research from political science, environmental sociology, governance, and the behavioral sciences has argued that the presence of certain variables increases the likelihood of cooperation in social dilemmas and collective action problems.[3] These include: the involvement of individuals that think in the long-term and see the maintenance of common resources as important for their own achievements; the availability of reliable information with minimal transaction costs related to its collection; open and frequent communication among stakeholders regarding costs and benefits; effective rule enforcement and provisions for forcing compliance (such as sanctions); and predictable and gradual changes to rules and enforcement when they occur.

Conversely, major changes in group composition, inflexible and homogeneous rules, rapid changes in technology, information failures between groups or generations, absence of transparency and legitimacy in governance procedures, dependence on external sources for resources or aid, and unchecked opportunistic or rent-seeking behavior seem to corrode effective governance and complicate cooperative efforts.

Climate change adaptation, of course, fits almost all of the criteria that make management difficult. Groups and stakeholders are constantly altering their composition and behavior in response to different climate threats. Rapid changes in both technology and climatic impacts complicate efforts at forecasting scenarios and calibrating optimal responses. Huge asymmetries in information exist between groups – consider the knowledge held by individuals at the insurance industry or the US Department of Defense about climate change risks compared to an illiterate farmer in Mali or a poor fisher in Bangladesh. Poor communities and countries are almost entirely dependent on international aid resources for assistance in strengthening adaptive capacity, and this book is replete with examples of rent-seeking and manipulative behavior where people utilize adaptation projects to advance their own agendas, even at the expense of others or the net common good.

How, then, *ought* adaptation projects be managed? And are there ways to design or promote them that can greatly minimize the chance of enclosure, exclusion, encroachment, and entrenchment from occurring? In this chapter, we propose a set of governance design principles for adaptation that focus on equity and distributive justice, inclusivity and procedural justice, information, accountability, and conflict resolution.[4] We then match these principles to five integrated policy mechanisms that we believe can address most of the damaging aspects of adaptation. These mechanisms include: taxing adaptation project externalities; critical stakeholder involvement; free prior informed consent; bonds and punitive liability coupled with adaptation inspection panels; and an international tribunal for handling grievances.[5] As we will see in the concluding section of the chapter, all five mechanisms need to be promoted synergistically rather than individually or piecemeal.

2 Five design principles

Because any policy action rests on certain fundamental assumptions, we thought it best to begin by presenting what we see as five core underlying design principles that help ensure adaptation projects are not only effective, but fair and just. These are:

1. equity, or proportional equivalence of costs and benefits that avoids having some users or adaptation actors getting all of the benefits at little cost;
2. inclusivity and due process, or the active involvement of a diverse number of stakeholders, usually brought about by participatory modes of decision-making or incorporation of local actors in the planning process, involving multiple institutions to deepen coverage and offer parallel systems of governance;
3. information, which includes the distribution of data about costs and benefits, the close monitoring of programs, and mechanisms that encourage feedback and the sharing of knowledge;
4. accountability, meaning that project sponsors bear some of the costs of governance themselves and are accountable for their actions, and also that graduated sanctions exist to enforce rules and penalize noncompliance; and
5. adaptability and arbitration, meaning that mechanisms are in place for conflict resolution and to respond to challenges and unforeseen events.

This part of the chapter investigates each of these five principles in turn; the next section applies them to the formulation of five policy mechanisms that planners and practitioners can actually implement.

2.1 Distributive justice and equity

The concept of equity is rooted in the idea that individuals should be treated fairly without regard to factors such as age, race, class, gender, or ethnicity. In practice, equity means that diverse actors enjoy a project's benefits; the spoils do not accrue to an elite minority. As philosopher Henry Shue put it, justice is about "not squeezing people for everything one can get out of them ... a commitment to justice includes a willingness to choose to accept less good terms than one could have achieved – to accept only agreements that are fair to others as well as oneself."[6]

Theories of philosophy, ethics, and jurisprudence often center on distributive justice, concerning themselves with how common goods (a fish stock) or common bads (pollution) are allocated among society.[7] Theorists call it "distributive" because it deals with three aspects of distribution:

1. What goods, such as wealth, power, respect, food, or clothing, are to be distributed?
2. Between what entities are they to be distributed (e.g., social groups, living or future generations, sentient beings, or all biodiversity)?
3. What is the proper mode of distribution (e.g., based on need, based on merit, based on property rights, or something else)?

Another way of framing distributive justice is to look at its scope, pattern, and currency of advantage. By "scope," distributive justice concerns who is identified as a legitimate provider or recipient of benefits and burdens. By "pattern," distributive justice investigates the level of benefit each subject of justice has a right to. By "currency of advantage," distributive justice concerns which accounts and assumptions policymakers and planners take into consideration when deciding how to allocate resources.[8]

Regarding the mode or allocation of distribution, many distinct approaches can be utilized. An "equality of rights" approach would spread resources thinly across an entire area, like spreading butter on a giant piece of toast. A "utilitarian" approach would seek to maximize overall happiness, balancing costs and benefits in the allocation of resources. A "democratic equality" approach would assign priority to the most disadvantaged, focusing on those already aggrieved or most vulnerable. A "proportional equality" approach would look to account for historical patterns in adaptation aid or proportional mitigation of greenhouse gases (e.g., where a country keeps current shares of total emissions, called grandfathering). A "causal responsibility" approach would assign those responsible for a problem – in this case, mostly large, industrialized countries emitting the bulk of historical emissions – the duty to fix it, something known also as the "polluter pays principle." A "merit-based" approach would distribute resources to those communities most active or involved, or with the best empirical performance.[9] This final approach is partly reflected in the notion of "common but differentiated responsibilities and respective capabilities," a cornerstone of the UNFCCC.[10] Legal scholar Burns H. Weston argues that there is no single standard of distributive justice; instead, distribution "can be based on equality, to everyone the same welfare, resources, or capabilities; priority, to each according to one's contribution or need; or sufficiency, to everyone enough to pursue one's aims and aspirations without major distress or dissatisfaction."[11]

How does one choose among this diverse array of options? The late philosopher John Rawls put forth the idea of a social contract. Rawls' idea was that people make a contract with one another and agree to give up the private use of force and other social ills, such as thievery or murder, in exchange for peace, security, and mutual advantage; a society exists when all people surrender some power before the law and greater good.[12] The idea also suggests that if we divest human beings from the artificial advantages they hold – wealth, rank, social class, education, musical ability, athleticism – they will agree to a contract of a certain sort.[13] Or, as Rawls wrote, we should take as a starting point that "reasonable principles of justice are those everyone would accept and agree to from a fair position."[14]

Rawls argued that two key principles must be accepted for a system of distributive justice to work: "equal basic liberties for all citizens, such as freedom of speech and religion" and social and economic benefits that offer security to the least well-off members of society.[15] Although egalitarian in nature, Rawls's position was not strictly egalitarian in the sense that all income, wealth, good, and services should be distributed at exactly the same levels to all members of a society.[16] Rawls instead supported the position that some inequality can actually benefit society as a whole, if it addresses a previous injustice, or targets a particularly vulnerable group of recipients.

When applied to climate change adaptation, distributive justice deals with who the recipients of justice are, what is distributed, and what principle determines that distribution. Options for recipients include those (currently) facing climate threats in any nation as well as the specter of climate change impacting future generations. Options for distribution include burdens, such as pollution, flood risks, and waste, as well as benefits such as access to the resilience, economic development, or employment an adaptation project might offer. Options for the principle of distribution include principles of equality, principles of a minimal standard, or preferential treatment to account for the special needs of recipients.[17]

Interestingly, beyond lofty moral theory, recent evidence has demonstrated that justice is key to individual and collective happiness as well as sound government. The psychologist Abraham Maslow, famous for his work on the "hierarchy of needs," wrote as early as the 1950s that people have an innate, psychological need for social justice. Social justice, for him, was a "precondition for the satisfaction of basic needs" such as self-esteem and warm food.[18] More recently, biologists at the University of California Los Angeles found that reactions to fairness and equity are "wired" into the brains of rats, consistent with the notion that "being treated fairly satisfies a basic need," almost similar in its effect on the brain as chocolate or money.[19] In parallel, scientists at Emory University have studied the behavior of Capuchin monkeys and noted that they possess a similar aversion to injustice and inequality as human beings, implying that a sense of justice is instinctual in nature.[20] In short: human beings are predisposed to expect, and demand, distributive justice.

2.2 Procedural justice and organizational multiplicity

Due process and fair decision-making procedures, especially the inclusion of rural actors, homeowners, and stakeholders (instead of just urban actors, commercial firms, and select groups) can enhance the effectiveness of adaptation projects and ensure adequate representation in how projects are implemented.[21] Generally, these protections center on three interrelated justice issues:

- Who gets to set rules and laws and which parties and interests are recognized in decision-making?
- By what process do they make such decisions?
- How impartial are the institutions, instruments, and objectives involved?

Unlike distributive justice, procedural theories of justice are oriented to process, including the fairness and transparency of decisions, the adequacy of legal protections, and the legitimacy and inclusivity of the institutions involved in decision-making.[22] Procedural justice deals with recognition (who is recognized), participation (who gets to participate and who has capacity to act), and power (how is power distributed in decision-making fora).[23]

Though procedural justice may strike some as dry and unimportant, fair procedures matter because they tend to promote better – more equitable but also more efficient and effective – outcomes, decisions, and investments.[24] Sociologist Claire Haggett compellingly argues that there are *pragmatic* reasons for ensuring due process.[25] For instance, public involvement can increase the likelihood of support, and therefore ultimate approval, for the siting of energy or climate projects. As she writes, "while fiscal regulations and subsidies, technical efficiency and political deliberations all affect the deployment of renewables, the stark fact remains that all of this matters little if there is no public support for a development."

There is also a strong case to be made that procedural justice is critical to the proper functioning of a deliberative democracy. In contrast to traditional models of democracy with established procedures for voting and representation to reach collective decisions, deliberative democracy emphasizes that collective decisions become legitimate through an open and reasoned dialogue among free and equal citizens.[26] Deliberative practices, however, can come saddled with potentially unequal procedural access. Deliberative processes, for instance, can be limited to a small group of citizens to ensure productivity. One study described this deliberative democracy paradox by noting that "deliberative decisions appear to be illegitimate for those left outside the forum, while bringing more than a few in would quickly turn the event into speech-making, not deliberation."[27] To counter this risk, deliberative theorists emphasize that participants in public deliberations need to act as political trustees representing the voices of those who are outside the deliberations.[28] Participatory processes thus require due process and procedural justice, attuned to more careful identification of participants based on stakeholder types and roles based on functions, geographical location, abilities, interest, and position in societal hierarchies.[29] The deliberations also need to counter the risk that participants may not participate on equal terms, and that some voices may not be heard even though they are physically in the room.[30]

Apart from contributing to the proper functioning of democracy, public involvement can also provide new knowledge, identify new goals, and contribute to the determination of what a good outcome even is or should be for a particular adaptation intervention. Physicist and philosopher Ian G. Barbour writes that "citizen participation is a source of information about public attitudes and perceptions of which assessors might otherwise be unaware" and that "later protests and opposition might be reduced if potential objectives were anticipated and alternatives were developed to take them into account."[31] Legal scholar Sanford Lewis adds that public involvement, open information, and deliberation can help to counter corporate cover-ups, false advertising campaigns, biased research, and the manipulation of the media.[32]

Furthermore, the involvement of multiple actors in adaptation decision-making processes can deepen protection and strengthen parallel systems of governance. Engaging different institutional partners and involving a diversity of important stakeholders in each project allows for the sharing of risks and invites organizational multiplicity, which can create "checks and balances" on actors.[33] In the literature on public policy and behavioral economics, this multiplicity is sometimes referred to as "polycentrism" or "polycentric approaches" to governance, when actors organize themselves (or are forced to organize) into multiple authorities on many scales simultaneously.[34] Two seminal studies showed how public policy problems related to street repair, policing, groundwater management, forestry, and irrigation were often best addressed through approaches that combined many units and many different scales, resulting in a diversity of arrangements and tools with differing actors, rather than a "monocentric" system of one actor on one scale.[35]

Polycentric approaches imply that the sharing of power between numerous scales of governance must be seamlessly managed, resulting in a "polycentricity" or "nestedness" that involves multiple authorities and overlapping jurisdictions.[36] Polycentric institutions are closely related to the "Type II" organizations depicted by Hooghe and Marks,[37] which are functionally defined (rather than geographically fixed) entities that possess varying boundaries and are established at the scale deemed best suited to deliver a particular service. The justification for polycentric approaches is that conventional forms of governance such as top-down centralized control, bottom-up decentralized control, and free market privatization have inescapable flaws in isolation.

Indeed, though far from definitive or absolute, emerging evidence suggests that polycentric approaches can encourage ideological plurality, promote dialogue, ensure redundancy, and enhance accountability when it comes to implementing energy or climate projects. Having multiple regulators means that different officials with distinct perspectives review a problem. This diversity of perspective produces a broader variety of potential solutions and facilitates experimentation. Having multiple interventions undertaken within the same system can also create opportunities to exchange results and observe experiences, and it can lead to the "adaptive management" of problems when they arise.[38] Under a system of polycentrism, if one level of government, industry, or civil society fails or defaults on a problem, other layers remain available to address the task. This creates a "safety net" that can ensure social problems are addressed.[39] Put another way, polycentrism can ensure that twice as many resources are allotted to address a particular problem. Lastly, polycentrism can improve accountability, autonomy, and participation. Multiple levels of government, the thinking goes, are harder to capture or corrupt than one. Moreover, polycentric forms of governance can create a greater sense of legitimacy and a high level of participation since they are situated closer to communities and are more responsive to local needs.[40]

2.3 Information and transparent monitoring

Our third principle relates to information and transparency: ensuring that all stakeholders are informed about an adaptation project, facilitating the close monitoring of programs, and providing mechanisms that encourage feedback and the sharing of knowledge.

New Zealand philosopher Gillian Brock argues the ability to be informed about what is happening to one's environment, possessing reliable, up-to-date, easily understood information, is an instrumental part of preserving fundamental liberties.[41] Principle 10 of the 1992 Rio Declaration on Environment and Development, adopted at the same time the UNFCCC was being signed, stated a global right to fair and transparent information as follows:

> At the national level, each individual shall have appropriate access to information concerning the environment that is held by public authorities, including information on hazardous materials and activities in their communities, and the opportunity to participate in decision-making processes. States shall facilitate and encourage public awareness and participation by making information widely available.[42]

Furthermore, the Aarhus Convention, which entered into force in October of 2001 and has since been signed by 40 European and Central Asian countries, clearly stipulates the necessity of information. Article 7 states that:

> Each Party shall make appropriate practical and/or other provisions for the public to participate during the preparation of plans and programs relating to the environment, within a transparent and fair framework, having provided the necessary information to the public.[43]

Similar statutes and claims supporting public access to information about energy and the environment appear in the 1992 Convention on Biological Diversity, the 1993 North American Agreement on Environmental Cooperation, the 1994 United Nations Convention to Combat Desertification, and a series of bilateral treaties.

Indeed, access to information and transparent frameworks for preserving that access have been known – under certain conditions – to reduce corruption and improve social stability. Kolstad and Wiig tested the correlation between corruption, defined as the abuse of public office for private gain, and transparency, defined as access to information, in 150 countries.[44] They found a "certain" correlation between lack of transparency and high levels of corruption. They noted, among other findings, that:

- A lack of transparency makes corruption less risky and more attractive;
- A lack of transparency makes it harder to use incentives to make public officials act honorably;
- A lack of transparency makes it hard to select the most honest and efficient people for public sector positions or as contract partners;
- A lack of transparency makes cooperation more difficult to sustain, and opportunistic rent-seeking more likely.

As just one example proving their point, in Uganda only 13 percent of grants for education actually reached schools throughout the 1990s; the rest were "captured" by the local government.[45] As soon as the Ugandan government began publishing the details of those grants in newspapers, 80 percent reached schools.[46] A similar study of 105 countries from 1960 to 2004 confirmed that "lack of transparency" can significantly and negatively impact economic growth. That study documented such a strong "causal relationship" across various sample sizes and time frames that it concluded that "the lack of transparency... is one of the primary reasons for the subsequent poor growth record for [a number of less developed] countries."[47]

General principles exist regarding how information ought to be created and disseminated, and transparency maintained. Sufficient relevant information must be offered in a way that is easily understood by non-experts. Sufficient time must be given for individuals to read and digest information and its implication. Mechanisms to incorporate feedback must be offered so that concerns can be addressed. The place and date of meetings must encourage maximum attendance and free exchange of views, including those opposed to a project.[48] Lastly, and critically, as participatory processes often reflect existing power structures, the processes must ensure that the most vulnerable are not only participating, but given political influence.[49]

Actual information engagement with the public can take a variety of forms. One type is "engagement as information provision." This is in line with the traditional "decide, announce, defend" mentality of informing people of plans that have been made, and distributing emails, leaflets, advertising, and providing exhibitions and displays. Another type is "engagement as consultation," which seeks to not only provide information to the public, but also to actively solicit feedback from them. Another type is "engagement as deliberation," where the public are not just permitted to discuss plans, but are thoroughly involved in their development. This shifts the public's role from participation to consultation, and though it is rarer, common examples include citizens' juries, interactive panels, and workshops.[50]

2.4 Accountability and responsibility

Our fourth principle is one of accountability and responsibility, holding adaptation project sponsors (or investors, or planners) accountable for the impacts of their projects and also stipulating graduated sanctions to enforce rules and penalize noncompliance.

One interesting foundation for this type of thinking is known as the "polluter pays principle:" whomever does the harm is responsible for undoing it. This is a matter of "corrective" action since when one group has engaged in wrongfully injuring another group, they should desist from their harmful actions and also compensate their victims for damages.[51] Such inequalities are "reversed" by imposing extra burdens on those responsible for inflicting and producing inequalities.[52]

Another core element of this principle is ensuring that strong penalties exist for those that break the rules. Aggressive non-compliance penalties have deterred non-compliant or illegal behavior in at least three other areas of energy and environmental law: air pollution, renewable energy policy, and endangered species. In the United States and Europe, sulfur dioxide allowance-trading programs, under which regulators impose automatic penalties ranging from $2,000 to $3,000 per ton (indexed to inflation) for each excess ton of SO_2 produced, have had compliance rates above 99.9 percent since penalties for noncompliance have been so severe.[53]

In the renewable energy credits market, the state of Texas provides one of the best examples of the success of setting high non-compliance penalties. In 1999, the Texas government required utilities to install 2,000 MW of new renewable capacity by 2009. Utilities failing to meet the standard had to pay the lesser of five cents per kilowatt-hour (expensive at that time) or 200 percent of the average price of renewable energy credits for each missing kWh. Because non-compliance penalties were set high above the cost of installing new renewable energy technologies, not a single utility failed to comply.[54]

Perhaps of most importance, protections of endangered species (the Endangered Species Act in the United States, or the Convention on International Trade in Endangered Species of Wild Fauna and Flora globally) are generally effective in industrialized countries because penalties for violation are so severe. In the United States, it is an offense to import or export, take, possess, sell, deliver, carry, transport, or ship illegal species, and the penalties for each violation can be a maximum fine of up to $50,000 or imprisonment for one year, and civil penalties of up to $25,000 per violation.[55] These severe penalties have been critical to the Act's effectiveness. Economists from Oregon State University and Tulane University used economic models to examine the statistical recovery rates for 225 vertebrate species protected under the ESA from 1990 to 2002, and they found that the ESA reduces the probability that a species will continue to decline once listed.[56] Using time-weighted analysis, conservation biologists from the Center for Biological Diversity and Cornell University compared population trends for 1,095 species covered under the ESA and found that "the longer species were listed, the more likely they were to be improving and the less likely to be declining," meaning that "ESA conservation measures act cumulatively over time."[57] In its broad assessment of the effectiveness of the Act, the Congressional Research Service noted that 41 percent of the roughly 1,700 listed species have improved or stabilized since listing, and that only seven of these species have gone extinct while protected under the Act – and five of these were later found to have been extinct at the time of listing.[58] After randomly selecting 107 out of 558 recovery plans (covering 1,049 species) under the ESA, the US Government Accountability Office noted in an earlier study an anticipated 100 percent success rate for *every* recovery plan with a time estimate – with seven species expected to fully recover by 2005, 19 more by 2010, and 12 more by 2015.[59] Clearly, strong penalties are a key part of effective environmental governance.

2.5 Adaptability and arbitration

Our final principle is one of adaptability and arbitration, meaning that mechanisms are in place for conflict resolution and to respond to challenges and unforeseen events.

The two usual processes that facilitate the resolution of conflicts are negotiation and arbitration, and both have much promise because they can give their own type of redress. Negotiation involves "direct dialogue between the parties to the grievance with the aim of resolving the grievance or dispute through mutual agreement," whereas arbitration is "the process by which neutral arbitrators selected by the parties to a dispute hear the positions of the parties, conduct some form of questioning or wider investigation and arrive at a judgment on the course of action to be taken in settling the grievance or dispute, often [but not necessarily], with binding effect on the parties."[60]

The reason negotiation and arbitration matter is because they can give redress, or meet a "right to remedy" or provide "access to remedies."[61] A right to remedy was first encapsulated in Article 8 of the UN Universal Declaration of Human Rights, adopted on December 10, 1948; it states that "everyone has the right to an effective remedy by the competent national tribunals for acts violating the fundamental rights granted him by the constitution or by law." The UN International Covenant on Civil and Political Rights (ICCPR), which became international law on March of 1976, also gives particular import to the rights of persons to seek remedy and redress; Article 2(3) suggests that access to remedies can encompass at least three dimensions: equal and effective access to due process (i.e., the courts or forms of negotiation and arbitration), adequate reparations for harm suffered, or access to information concerning the violators. Most recently, in 2011 the Human Rights Council of the United Nations adopted a set of principles urging states to protect the fundamental freedoms and human rights of their citizens, and urging businesses to adhere to all international and local laws, and arguing that when either fails to do so, "the need for rights and obligations [should] be matched to appropriate and effective remedies when breached."[62] Amnesty International adds that "all victims of human rights violations have a right to an effective remedy" and that "this right lies at the very core of international human rights law," stemming from "a general principle…that every breach gives rise to an obligation to provide a remedy."[63]

3 Five policy mechanisms and best practices

Confucius is believed to have once written that "one that merely knows right principles is not equal to one that loves them."[64] Or, as distinguished ethicist Peter Singer wryly notes, with justice "the proof of the pudding lies in the eating."[65] Principles mean little unless they are *acted* upon. With an eye for how our five principles can inform the practice and better management of adaptation projects, we offer five policy mechanisms in this section of the chapter: taxation of externalities, critical stakeholder involvement, free prior informed consent, bonds and inspection panels, and an international tribunal for conflict resolution.

3.1 Account for and tax adaptation's externalities

Our first policy prescription is to incorporate the full price of adaptation (including as many of its externalities as can be priced) into project budgets, perhaps through an "adaptation tax" similar to a carbon tax.

To make sure such externalities are tracked and accounted for (as best as they can be), we propose that the UNFCCC mandate its Standing Committee on Finance to consider the externalities of adaptation in its directives for measurement, reporting and verification of support provided through, for example, the Adaptation Fund and the Green Climate Fund.[66] Whatever the price for storm surge barriers, it must include the costs of degraded seascapes, the price of decommissioning, and the risk of a severe accident. The cost of mangrove afforestation must include money for planting trees as well as compensation for those forcibly relocated from the conservation area and retraining for farmers who have lost their jobs. The cost of disaster relief following a hurricane must include not only physical reconstruction and rehabilitation, but also funding to defray the social impacts facing migratory refugees and sensitivity to environmental recovery and bioremediation. To be fair, not all things can be quantified; not all of those that can be are consistent; comparisons seldom assess costs and benefits across technologies; and most methods suffer from methodological shortcomings.

That said, there has been some synthetic scholarship looking comprehensively at the cost of externalities across a wide range of social and environmental impacts. A hypothetical list of potential externalities includes:

- Catastrophic risks such as sea wall breaches or the collapse of forests due to pests;
- An increased probability of social conflict due to natural resource extraction for adaptation projects;
- Public health issues and chronic disease, morbidity, and mortality;
- Worker exposure to toxic substances and occupational accidents and hazards;
- Public deaths and injuries due to construction equipment or the building and climate-proofing of infrastructure;
- Direct land use and the destruction of land including acid drainage and resettlement;
- Degradation of cultural icons such as national parks, recreational opportunities, or activities such as fishing or swimming;
- Incidence of noise and reduced amenity, aesthetics, and visibility.

The costs of these externalities can be factored into the price of adaptation projects, filtered into the price of carbon credits, or, similar to carbon dioxide in some European countries, reflected in environmental taxes.[67] What we are suggesting is that polluters start paying their fair share of enhanced resilience and improved adaptive capacity, and environmental victims start recovering for damages inflicted on communities and ecosystems around the world – that the "hidden" costs of adaptation become factored into its market price.

Taxing adaptation in this way brings three advantages beyond simply raising money for projects. First, pricing externalities would incentivize adaptation actors to minimize their damage. The current economic system facilities the opposite, the maximization of externalities. The more an actor can externalize its cost of doing business, the greater return on capital it will receive in the short term.[68] Second, pricing externalities brings issues of utility and justice to the center of the marketplace. As they stand now, externalities are wholly undemocratic and exclusionary. The most affected parties – often the poor or disenfranchised – are under-represented in the marketplace, and have external costs imposed upon them.[69] Yet on a per capita basis, lower income households contribute less to greenhouse gas emissions and climate change. Pricing externalities would reveal this inequality. Third, pricing externalities properly align price signals. It is not that we do not pay for adaptation's externalities as a society; we do through hospital admissions, lost worker days, higher taxes, blighted landscapes, death, pain, and suffering. Society and nature "pay" eventually, it is just that those costs do not show up on our monthly bills. Put simply: adaptation's price does not match its cost.[70] Including externalities ensures that it starts to.

Such taxes do not necessarily need to come from rural areas or vulnerable communities implementing adaptation projects. To ensure that adaptation and climate funding is fair, the ethicist Paul Baer and his colleagues have developed a principled framework known as Greenhouse Development Rights (GDR).[71] The GDR framework is based on the twin notions of capacity and responsibility reflecting the principle of Common but Differentiated Responsibilities and Respective Capabilities. It creates an emissions budget, and then divides the allowances by income according to individuals, making it a "burden sharing" tool. The notion of *capacity* is rooted in income for an individual adjusted for power purchasing parity over a development threshold. The GDR framework therefore distinguishes between necessities such as food and water and luxuries such as yachts and couture dresses. Capacity, in this way, means the ability to reduce one's consumption, and thus emissions, without sacrificing necessities; Bauer and colleagues calculated necessities at about $7,500 per person. Bauer and his colleagues define *responsibility* as one's contribution to the problem of climate change, adhering to the classic "polluter pays principle." Bauer and colleagues based this on per capita emissions of carbon dioxide from fossil fuel consumption. They then plugged all of this into a Responsibility and Capacity Indicator, RCI, shown in Table 6.1, which they projected out to 2010.

Essentially, Table 6.1 argues that people within the United States and European Union in 2010 should be responsible for 75 percent of the resources needed to address climate change, even though these regions hold less than 19 percent of the global population; and that they ought to start paying an average of $447 per taxpayer to help other countries build their adaptive capacity and mitigate their own emissions. The GDR framework thus ensures that developing countries receive necessary money for adaptation projects, while the higher costs and taxes for rich countries both fund that development and create incentives to limit and lower developed countries' consumption, and thus emissions.[72] As discussed in Chapter 5 on the UNFCCC, we know that historical responsibility as a principle has proven very difficult, if not impossible, to operationalize in distributing commitments within global climate negotiations. Another alternative would be to build on the convention's conceptual recognition of the principle, but raise contributions to adaptation finance based on capacity to pay in terms of GDP per capita.[73]

Table 6.1 Characteristics of a GDR framework based on responsibility and capacity, 2010

	Total Income (billion $ PPP)	Total Capacity (billion $ PPP)	Global RCI (%)	Per Taxpayer Bill ($ PPP)
United States	14,226	11,909	31.8	733
European Union-27	14,845	11,192	24.8	412
Russia	2,173	1,166	3.9	228
Brazil	1,891	926	1.8	166
China	8,567	2,345	6.6	124
India	3,541	321	0.8	65
Mexico	1,368	730	1.7	199
South Africa	502	257	0.9	320
Least Developed Countries	1,018	36	0.07	57
Annex I	39,357	30,112	74.6	447
Non Annex I	29,400	9,973	25.4	169
World	68,757	40,085	100	315

Source: Modified from Baer, P., Athanasiou, T., & Kartha, S. (2007). *The Right to Development in a Carbon Constrained World: The Greenhouse Development Rights Framework*. Heinrich Boll Foundation.

3.2 Critically involve stakeholders

Complementing better pricing of adaptation's externalities, more critical and broader stakeholder assessment is needed to ensure procedural justice and diversity. One mechanism for engendering this is critical stakeholder analysis, a framework for understanding allies and competitors whose involvement can impact the success or failure of a project or policy. While the word "stakeholder" first appeared in 1708 to refer to a person that held a stake or stakes in a bet or game of chance, its modern conception refers to a group or individual that can affect or is affected by an organization or event (or in this case, a particular adaptation effort or project). Critical stakeholder analysis identifies relevant stakeholders for a given project or policy, maps out their relative power, influence, and interests, and assesses the broader context in which they interact.[74] It involves looking at not only the type of actors and roles of participants involved in a given situation, but also evaluates their relative power and influence, resources, and the external social and political environment that they inhabit.[75]

Critical stakeholder analysis yields three important benefits. It can jumpstart dialogue and facilitate discussions among previously disconnected actors, making it an important component of democratic decision-making and in revealing power asymmetries between stakeholders.[76] In addition, the process of identifying stakeholder interests can promote common understanding of key agendas and help incentivize collaboration. Conversely, it can identify zero sum tradeoffs and incommensurable or irresolvable views among stakeholders that must be resolved for consensus to occur.[77] Finally, by making some stakeholders and their power relations more visible, critical stakeholder analysis can improve social responsibility and force desirable change. Numerous examples abound of organizations and stakeholders forced to alter their practices in response to stakeholder analysis and participation, especially those related to the destruction of the environment and human rights abuses.[78]

By collecting and analyzing data on stakeholders, one can develop an understanding of – and possibly identify opportunities that influence – how adaptation decisions are made in a particular context. According to Dunn, stakeholder analysis involves four steps:[79]

- Identify relevant stakeholders to include for analysis;
- Assess each stakeholder's objectives, interests and beliefs;
- Characterize stakeholder resources; and
- Elucidate the strategies and venues stakeholders use to achieve their objectives.

The next logical question is, of course, who represents the public, or who counts as a stakeholder? In most cases, typical stakeholders and representatives of the public would include:

- Local people (individuals) and communities (for example, villages) likely to be affected by a project;
- Traditional leaders or representatives on community-level bodies such as ward councils who can be consulted to obtain a community viewpoint, as well as non-resident social groups who might use local resources such as pastoralists;
- Selected social categories, for example women, the elderly, or the poorest in society;
- Religious leaders;
- Politicians;
- NGOs and voluntary organizations such as local community development or resource user groups, gender-based groups, labor unions, and cooperatives;
- Private sector bodies such as professional societies, trade associations and chambers of commerce;
- Media (newspapers, radio, television); and
- National and local government ministries, departments and statutory agencies whose responsibilities include areas and sectors likely to be affected (such as health, natural resources, and land use).

Consultation and involvement with these stakeholders can take a variety of forms. It can include public meetings without restrictions on access and attendance, or advisory panels chosen to represent stakeholder groups. It can include open houses – a staffed facility in an accessible location that contains information on a proposed project – or interviews and surveys with selected community representatives. It can also include some of the techniques related to free, prior informed consent in the next section.

3.3 Adhere to free prior informed consent

To meet our principle of information and transparency, we propose that all adaptation projects follow the process of free prior informed consent – FPIC – which refers to "a consultative process whereby a potentially affected community engages in an open and informed dialogue with individuals or other persons interested in pursuing activities in the area or areas occupied or traditionally used by the affected community."[80] The Center for People and Forests describes it as "the establishment of conditions under which people exercise their fundamental right to negotiate the terms of externally imposed policies, programs, and activities that directly affect their livelihoods or wellbeing, and to give or withhold their consent to them,"[81] making it distinct from mere negotiations and consultation.

Each of the four words in "FPIC" is meaningful.[82] "Free" implies that no coercion, intimidation, or manipulation has occurred; that potentially affected people freely offer their consent. This should ensure that the least powerful, often silent voices in participatory processes, become more actively consulted. "Prior" implies that consent has been sought sufficiently in advance of any meaningful decision to proceed with a project, before things like financing or impact assessments begin. "Informed" means that information about the project is provided, covering: its nature, size, pace, reversibility, and scope; expected costs and benefits; the locality of areas to be affected; personnel and revenues likely to be involved; and procedures for resolving conflicts, should they occur. In other words, affected communities must understand their own rights, and the true implications of the project as best as they can be known, so that opponents and proponents can negotiate with equality of information. Lastly, "consent" means "harmonious, voluntary agreement with the measures designed to make the proposed project acceptable."[83] It does not necessarily mean absolute consensus – a majority of 51 percent suffices some of the time – and it does not always mean direct involvement; consensus can occur through representatives, through plebiscites (direct single issue votes), and referenda. It is, however, distinct from consultation – the act of merely discussing a project with a community – because it gives communities the ability to say "no."[84] We think prior consent should be the norm. It is possible that there may be occasions when other considerations could require a project to go ahead for the greater public good, but these should be under exceptional circumstances, and fully justified.

Though the concept of FPIC is new compared to other instruments of international law, it has been honed under stress by a flurry of agreements. One of its earliest codifications was the Nuremburg Code of 1947 concerning the conditions under which medical experiments could be carried out on human beings. Various iterations of consent have been stated by the 1966 International Bill of Rights, 1966 International Covenant on Economic, Social and Cultural Rights, and 1966 International Covenant on Political and Civil Rights, which all entered into force in 1976. The 1989 International Labor Organizations' Indigenous and Tribal Peoples Convention states that indigenous peoples must be consulted about development projects on their territories, and the United Nations Food and Agricultural Organization established a 1989 "Code of Conduct" making consent mandatory. The 1992 Convention on Biological Biodiversity acknowledges that indigenous knowledge may only be used with prior approval and requires governments to protect indigenous cultures. Annexes to both the 1994 UNFCCC and 2009 REDD+ set safeguards detailing how vulnerable communities ought to be protected from energy projects. The 1989 Basel Convention, 1998 Rotterdam Convention on Free Prior Informed Consent, 2001 Stockholm Convention on Persistent Organic Pollutants, World Bank's Forest Carbon Partnership Facility and Forest Investment Program, Voluntary Carbon Standard practiced by some institutions trading credits under the 1998 Kyoto Protocol, and the Community Climate Biodiversity Alliance (a consortium of international NGOs) also refer to FPIC as an important "right."[85]

3.4 Implement punitive bonds and inspection panels

To hedge against encroachment, and to meet our principle of accountability and responsibility, we propose the creation of strict bonds and liability standards for adaptation projects, as well as the adoption of truth commissions and inspection panels.

One way of ensuring responsibility is to mandate it as part of a punitive bonding system. In the words of economists Laura Cornwell and Robert Costanza, environmental bonds "require those seeking to use society's resources to post a bond in advance of any potentially, environmentally damaging activity...the most important aspect of the bond amount is that it provides the incentive to behave in an environmentally responsible manner."[86] Such a bond, kept in an interest-bearing escrow account by an independent third party, can be tied to either the value of the resources involved or the likely value of environmental or social damage expected over the course of an adaptation project. If resource users or adaptation managers can demonstrate that the damages to society or the environment were less than the amount of the bond, this difference and a portion of earned interest can be refunded. The bonding system thus ensures that the financial resources available for protecting the environment, or local communities, are roughly equal to the potential harm arising from the design and implementation of an adaptation project. If and when damages occur, the bond ensures that funds exist to rehabilitate or repair the environment and compensate injured parties. This, in practice, is more like environmental "bail" than a traditional interest-bearing "bond" since its punitive value is greater than its investment value.

Though the idea of environmental bonding may sound radical, it already exists in various forms. Refundable deposits on glass bottles, for example, aim to encourage bottle users to dispose of the commodity in the most desirable way (by recycling) and to avoid its disposal in the least desirable way (as litter). By demanding that consumers pay in advance for the costs they might inflict on society, such deposits reverse the usual presumption of innocence over guilt as applied to environmental damages. Another precedent for environmental bonds is the producer-paid performance bonds frequently required for construction work. The 1935 Miller Act in the United States requires that contractors performing construction for the federal government secure performance bonds, a contractual guarantee that the principal (the entity which is doing the work or providing the service) will perform in a designated way. The National Environmental Policy Act of 1969 further required the owners and operators of deep-water ports to "establish and maintain evidence of financial responsibility sufficient to meet the maximum liability to which the responsible party could be subject to."[87] The Surface Mining Control and Reclamation Act of 1977 stipulated the use of performance bonds to guarantee reclamation of mining sites for coal and other minerals. Moreover, roughly a dozen states in the United States allow for the utilization of bonding instruments for environmental protection, and the federal government also requires the posting of a bond guaranteeing the complete and timely plugging of oil and gas leases on public lands as well as reclamation and restoration of any lands or surface waters affected by the lease operation. Additionally, the Nuclear Regulatory Commission stipulates regulations similar to bonds for the decommissioning of nuclear facilities and the storage and disposal of nuclear waste.[88]

Environmental bonds for adaptation would work in a similar manner by providing a contractual guarantee that the principal of the bond performs in an environmentally benign, and socially responsible, manner; they would be levied for the current best estimate of the largest potential damages. Whereas insurance protects companies against loss, environmental bonds guarantee the fulfillment of an obligation. Responsibility for honoring the bond's conditions would fall both to the surety company and the principal.

In terms of accountability, we propose the creation of adaptation truth commissions or inspection panels. Though they have not been applied to energy or climate issues before, truth commissions have been successful at holding people accountable for human rights violations in places such as Rwanda and South Africa.[89] Formal inspection panels offer another tool to document and remedy abuses and grievances. The exemplar here is the World Bank's Inspection Panel, which provides a forum for people to directly raise their concerns about *any* of the World Bank's projects to an independent group of experts. The World Bank created its Inspection Panel in 1993 after an independent review of the its support for the Sardar Sarovar Dam on the Narmada River in India concluded that it "clearly violated" Bank policies and had "devastating" human and environmental consequences, claims that were quickly substantiated.[90] The World Bank thus established an Inspection Panel consisting of at least three members nominated by the President of the Bank and approved by the board of directors, financed by the Bank, but serving independently for five-year terms – with members disqualified if they had worked for the Bank for at least two years and unable to work for the World Bank after serving on the Panel.

Two or more local people can bring a claim to the Inspection Panel, asking for an objective analysis of the World Bank's role in a project and the degree to which it violated or failed to comply with any of its policies and procedures. Claims are typically resolved within one year, though some have taken as little as a few months.[91] Between 1993 and 2008, the Inspection Panel heard 52 complaints, finding in favor of Requestors 45 times, and Requestors have succeeded at blocking the construction of socially and environmentally destructive dams in Nepal, coalmines in India, sewage treatment facilities in the Philippines, and various attempts to unfairly resettle local people.[92] So far, such Inspection Panels have taken hold at most multilateral development banks and a few other institutions, but we believe that international bodies such as the United Nations and private companies, among other actors, should begin to create and support their own inspection panels, especially in the sphere of climate change adaptation.

3.5 Create an international tribunal for redress

Our final mechanism is to create an international Tribunal for Adaptation Justice and Conflict Resolution, to handle grievances and to give redress. This could be modeled in part on a proposal from Lawyers for Better Business for a tribunal dealing with human rights abuses committed by businesses.[93]

In terms of function, the Tribunal would provide reparations to those harmed by adaptation projects.[94] It would award various types of damages similar to how US courts award compensatory damages in tort cases. We anticipate that the Tribunal would recognize at least five distinct forms of reparation:

- Restitution, broadly defined as restoring a person or community to their previous state of liberty, enjoyment, financial security, cohesion, etc.;
- Compensation, monetary awards to account for physical or mental harm, lost opportunities or employment, material damage, and costs required for legal assistance;
- Rehabilitation, encompassing medical and psychological care needed by victims;
- Satisfaction, referring to measures which verify facts and lead to full and public disclosure of the truth, a public apology including acknowledgment of the facts and acceptance of responsibility, and sanctions against perpetrators; and
- Guarantees of non-repetition, changes in law, policy, or oversight that provide a reliable measure of protection against further abuses.

While it would be up to the Tribunal to decide which forms of reparation are most appropriate for which cases, we would expect that some of the more severe cases involving irreversible impacts might necessitate many or even all of the five forms.

In terms of structure, the Tribunal would be unaffiliated with any particular government, and maintain a roster of judges and jurists highly familiar with human rights law, international law, environmental law, and climate policy. The Tribunal would apply tort law and legal principles involving climate change around the world, independent of the location or type of dispute. A secretariat could support the Tribunal, with funding from the UNFCCC (such as the Green Climate Fund), the adaptation tax or punitive bonds discussed above, a rights-based financial mechanism (such as the GDR framework) discussed above, or – as a last option – a range of donor countries. It could sit as part of the Rome Statute of the International Criminal Court and also reside in Rotterdam. The Tribunal could deploy arbitration and mediation, both being methods of alternative dispute resolution, and it could receive information from the truth commissions and inspection panels already proposed. To facilitate ease of access (especially among the rural or poor), it could, lastly, operate according to simplified discovery procedures, less formal pleadings, and flexible cross-examination and written decisions. It would, for instance, switch the burden of proof and establish a rebuttable presumption that adaptation actors are responsible for proving that they did not harm a given victim, rather than leaving the burden of proof where it currently rests, on claimants. The Tribunal could lengthen the statute of limitations beyond the usually stipulated ten years (such as under the Torture Victim Protection Act in the United States) – this would be elemental in covering adaptation projects many decades old, such as the Eastern Scheldt storm surge barrier and dikes in Bangladesh, or even the Hurricane Katrina recovery efforts, which had their ten year anniversary in 2015. The Tribunal could eliminate the "loser pays" statutes that often require each side in litigation to pay attorney fees, which can act as a major barrier to entry for poorer clients. It could also restrict the granting of amnesties to those who have committed abuses, and provide funds for the building of capacity for claimants and attorneys, so that claimants can have an easier time collecting evidence and testimony and special prosecutors can gain experience and expertise related to the investigation and prosecution of crimes and human rights abuses related to adaptation.

4 Uncomfortable conclusions and implications

Even concerning our recommendations, we present three conclusions with perhaps uncomfortable implications for the design and practice of climate change adaptation projects.

First, all five of our principles and policy mechanisms are needed as part of a concerted package; implementing one or two by themselves will be insufficient to offset or minimize the processes of enclosure, exclusion, encroachment, and entrenchment. For example, principles of equity and social justice, based on so-called notions of distributive justice, differ from due process and procedural justice, related to principles cutting across inclusivity and information. Distributive justice is about how adaptation projects disseminate their costs and benefits, whereas procedural justice is about free prior informed consent for adaptation projects, representation in adaptation decision-making, ensuring the voices of the most vulnerable are given political weight, and facilitating access to high quality information about adaptation projects. Our aspects of accountability and arbitration cut across both dimensions, since they touch upon the distribution of goods and bads as well as the fairness of the decision-making process. However, promoting one form of justice without the other might make both ineffective. As Paavola, Adger, and Huq put it, "redistribution without empowerment can be short-lived, and empowerment without redistribution can be an insult."[95] That is, proper distribution of outcomes – distributive justice, and impartiality in decision-making – procedural justice, must be part of a unified package. Or, as one philosopher explained: "my utility may not only depend on what I get but on the manner in which I get it. That is, my utility may not only depend on the consequences of policy but on the policy itself."[96]

More pragmatically, failure to holistically implement our five policy mechanisms can still give rise to problematic projects. Failing to tax adaptation's externalities leaves little funds available for an international tribunal with the teeth and power to offer meaningful reparations. Failing to sponsor truth commissions and inspection panels also makes it difficult for those hurt by adaptation projects to document damages that they can take to the tribunal or use in establishing damages under our proposal for punitive bonds. Critically involving stakeholders but not mandating that they adhere to free prior informed consent might only enable the most powerful actors to act as hegemons and roll over others in the pursuit of their interests during stakeholder negotiations. In this way, each mechanism is necessary but insufficient by itself to best manage the political economy processes of adaptation.

Second, though this chapter has mapped out five design principles and corresponding policy mechanisms, it only gets us half of the way toward more inclusive, just, and equitable adaptation projects. That is because it focuses on policy solutions but not politics – the social or political conditions that could actually bring about the five changes we suggest. To be sure, there are immense obstacles and entrenched interests in the current system that would fight – perhaps bitterly and fiercely – to maintain the status quo. Indeed, one could even say that change will need to match the extent to which the processes of enclosure, exclusion, encroachment, and entrenchment operate, because every stakeholder benefitting from those processes has a strong reason to oppose reform. The implication here is that while all five mechanisms are certainly needed simultaneously, they will need strong support from a leader or set of leaders, or public backing to become politically acceptable.

Third, many of our policy mechanisms require political and economic sacrifices that elites might not be willing to make. Taxes on adaptation externalities will fall almost exclusively on the wealthiest nations and individuals in the world. The critical involvement of diverse stakeholders will dilute the power and influence of elites. Adhering to free prior informed consent will undermine neo-colonial land grabs that enrich a few at the expense of many. Punitive bonds and inspection panels will require up-front resource outlays and increase investment risks, saddling the wealthy with new debt and uncertainty. A tribunal for redress – if able to avoid the pitfalls of many current international justice mechanisms[97] – would prove effective only if it redistributed resources and power to the victims of the political economy processes we describe. In sum, our proposals test the limits of our collective capacity to set a new course for human relations and climate change adaptation – though this is also perhaps why they are so sorely needed.

7
Insights from Political Economy for Adaptation Policy and Practice

1 Introduction

Environmental studies professor David W. Orr once wrote about a remarkable experiment involving two different groups of kittens, raised in rooms that differed only in the color of their walls.[1] One group was raised in a room painted with horizontal lines, the other with vertical lines. After several weeks, the kittens were moved from one room to the other. Despite the fact that both environments were identical with the exception of the lines on the walls, both groups suffered severe adjustment problems, including higher mortality rates. The implications of the study concerning perception and adaptability to the environment are interesting, but also telling in a human context. Orr's research prompts us to ask: if we are forced to adapt to much more serious situations, will humans experience similar degrees of coping difficulties? More importantly, if the political economy aspects we identify in this book are correct, will some malevolent actors *create* disorientation by design, utilizing chaos and confusion to hide the underlying processes endowing them with wealth and power?

In this book, we propose that climate change adaptation projects can, at times, propagate four interconnected political economy processes that operate systemically. Enclosure refers to when public assets become privatized or private actors extend their reach and autonomy into the public sphere. Exclusion refers to when adaptation actors marginalize particular groups in decision-making fora. Encroachment refers to when adaptation projects damage the environment, notably biodiversity conservation zones or fragile ecosystems. Entrenchment refers to when adaptation projects worsen social or economic inequality. These four processes suggest the human responses to climate change can become intertwined with the perpetuation of aggressive capitalism, imperialism, hegemony, and greed. More specifically, through four case studies we have shown how adaptation projects can consolidate local elite control (Bangladesh), damage ecosystems and socialize flood risk (the Netherlands), exacerbate poverty, racism, sexism, and classism (the United States), and weaken the efficacy of global climate diplomacy (the UNFCCC).

This chapter presents five core conclusions resulting from our analysis. The first conclusion is that the processes of enclosure, exclusion, encroachment, and entrenchment, presented across four case studies examined in the book, can subvert the intended goals and desirable attributes of adaptation projects. The second conclusion is that conflict and struggle are inescapable in adaptation projects, meaning they will always result in distinct winners and losers. As such, while many and perhaps even most adaptation actors have heroic intentions, not all actors or stakeholders are benevolent. The third conclusion is that even though they are often conceived as local or subnational processes, our analysis suggests that adaptation projects are enmeshed in complex global relationships. The fourth conclusion is that five interconnected policy options need to be implemented to hold the political economy processes of adaptation in check. When these are adhered to, adaptation projects need not be rejected, rather they can be calibrated to ensure that they reach their goals in the most equitable ways possible. Our fifth conclusion is that our analysis, though we believe it to be valid, also points the way for future research to strengthen our understanding of both adaptation and political economy.

2 Political economy can subvert the goals of adaptation projects

This conclusion is the most direct and simple: the political economy of adaptation, namely the processes of enclosure, exclusion, encroachment, and entrenchment, can distort the goals and effects of adaptation projects. All four of our case studies exhibited all four of our political economy processes, which we summarize in Table 7.1. Interestingly, the political economy challenges involving adaptation are not relegated to developing countries such as Bangladesh or those receiving aid and technology from the UNFCCC, they occur even in countries with strong regulatory regimes and standards such as the Netherlands and the United States.

Table 7.1 Summary of political economy of adaptation case studies

Process	Case Study	Explanation
Enclosure	Bangladesh	Powerful landlords (*jotdars*) utilize a neo-feudal system of land grabbing and appropriate public lands to consolidate power and concentrate wealth
	Netherlands	National planners use flood control as a justification to extend their control over local collective water organizations and councils and continue on a path of professionalization which marginalizes those outside of formal institutions
	United States	Public Katrina recovery resources get directed primarily to help large corporations (casinos, cruise line ships, port operators), and homeowners, private actors use recovery to facilitate their own housing development plans or the capture of public school buildings
	United Nations	Decision-makers and diplomats representing powerful countries or coalitions use the UNFCCC process to establish an intricate web of bureaucracy and authority on the topic of climate change
Exclusion	Bangladesh	Representatives from vulnerable community groups, their professional associations, and civil society organizations are excluded from national climate policy fora, and are unable to use the courts to seek redress
	Netherlands	Construction contracts for the Dutch Delta Works get directed to large national firms, the *Rijkswaterstaat* exerts a command and control management strategy that limits access to information and stifles political debate over at least two decades
	United States	Businesses take advantage of low-wage undocumented workers in reconstruction work, excluding others; minority evacuees are more likely to face rent hikes or eviction; women and disaster front-line volunteers are excluded from planning discussions
	United Nations	Rules of procedure tend to benefit powerful countries and coalitions at the expense and marginalization of poorer countries

Encroachment	Bangladesh	People become displaced or lose livelihood opportunities in the face of climate-infrastructure, forest conservation areas, waterways, and aquaculture sites
	Netherlands	Marsh vegetation, mussels, snails, harbor porpoises and especially birds see their habitat and spawning ground degraded by the Delta Works, especially across the Eastern Scheldt Estuary
	United States	Hurricane Katrina reconstruction involves infrastructure repair and dredging done in conditions of relaxed environmental standards
	United Nations	Some adaptation projects incentivized or supported through UNFCCC mechanisms degrade the environment or interfere with established land use practices
Entrenchment	Bangladesh	Moneylenders and community leaders take advantage of disaster victims by forcing them to sell assets at low prices or indoctrinate them into a patronage system managed by the *matabbar* and *talukdar*
	Netherlands	The Dutch Delta Works entrenches flood risk among some peripheral communities
	United States	Levy rehabilitation centers on protecting wealthier geographic parts of New Orleans; poor, minority, female evacuees less likely to have access to transport, more likely to face forcible resettlement or remain in temporary shelters; private developers use the disaster as an opportunity to promote their own agendas
	United Nations	An inability to fund pledged commitments and restrictions on technology transfer keep many developing countries, in particular Least Developed Countries, entrenched in macro-economic inequality and technological inferiority

Source: Authors.

For example, all four case studies demonstrated the process of enclosure, the capture of public resources or expansion of private roles into the public sector. In Bangladesh, coastal adaptation projects have been used to "grab" *khas* and *char* lands designated for poor communities or public use. In the Netherlands, the Dutch Delta Works enabled national planners, notably those at the *Rijkswaterstaat*, to strengthen their control over the local collective water organizations and councils that had been making decisions about water for almost a thousand years. In the United States, Hurricane Katrina recovery efforts securitized and militarized disaster relief through agencies such as the National Guard, consolidated corporate control and profiteering over publicly available disaster relief aid, and enabled private actors to procure public housing projects and local schools. The UNFCCC continues to facilitate enclosure by marginalizing and eroding the autonomy of some public and even private stakeholders and groups in the COP process.

All four case studies had elements of exclusionary forms of planning and implementation. In Bangladesh, laypersons and NGOs charged with protecting indigenous communities have been unable to participate in setting the country's goals as enshrined in Bangladesh's National Adaptation Program of Action (NAPA). In the Netherlands, the size and complexity of the Dutch Delta Works meant mostly large firms participated in construction; the objections of fishers, yacht owners, local politicians, and some environmentalists were ignored for more than a decade. Katrina recovery in the United States excluded large segments of evacuees from motorized transportation and accessing more affluent neighborhoods, favored cheap, undocumented laborers over legal employees in construction work, and delegated most relief funding for middle-class homeowners. Deliberations within the UNFCCC consistently steer international negotiations in the direction of certain interested, dominant parties.

All four cases facilitated encroachment – negative land use or environmental impacts. Dikes and other sea protection infrastructure across Bangladesh have resulted in the clearing of forests and farms. In the Netherlands, the construction of the Delta Works degraded ecosystems known for their unique wetlands status and bird sanctuary, converting tidal areas rich in biodiversity into stagnant lakes with greatly reduced water quality. Hurricane Katrina relief depended in large part upon massive reconstruction efforts utilizing heavy machinery and pumps, waterway dredging, a relaxation of environmental standards, and a concentration of debris and pollution among poorer communities. UNFCCC-backed adaptation projects can necessitate diverting water from local residents, interfering with biodiversity conservation zones, or displacing communities from their land.

All four cases, lastly, exhibited entrenchment, the aggravation of some form of inequality. In Bangladesh, household coping strategies fall prey to manipulation and predation by local criminals. In the Netherlands, the Delta Works have shifted flood risks to other areas and deepened or entrenched them in some particular rural provinces. In the United States, Katrina recovery efforts have only exacerbated pre-existing class, racial, and gender inequalities, forcing predominantly poor, African American, and female-headed households to struggle through lives of "permanent temporariness" and "Katrina fatigue." Entrenchment within the UNFCCC involves an obvious lack of fair (and promised) funding for adaptation as well as limited transfers of adaptation knowledge and technology.

Although we maintain that these four political economy processes are analytically distinct, they also often interact in complex, mutually constitutive ways. In Bangladesh, the local, almost feudal-like predatory processes that entrench social dependence on bandits and criminal gangs are also reinforced via a de facto exclusion from the judicial system or other forms of redress – entrenchment mutually begot exclusion. In the Netherlands, decades of enclosure and exclusion by the *Rijkswaterstaat* over the grand Delta Works plan, and an obsession with national defense and safety, made government actors all but blind to its staggering environmental implications – enclosure and exclusion begot encroachment. In Hurricane Katrina recovery efforts, the funds available for levy repair undertaken by the US Army Corps of Engineers were unavailable because large sums of disaster relief want to corporations who had enclosed upon the resources, resulting in unequal protection and a consequent increase in housing risk and rise in insurance prices – enclosure begot entrenchment. Within the UNFCCC, the enclosure of authority over climate change policy and exclusion of marginalized developing countries leads to patterns of adaptation that are less responsive to local needs – enclosure begot encroachment and entrenchment.

These examples illustrate the interstitial nature of the political economy processes; they often result in self-fulfilling feedback loops that make the political economy of adaptation systematic and interactive.

3 Adaptation will always create winners and losers

Our next conclusion is that conflict and struggle are part and parcel of the process of adaptation. No matter how noble the intentions of planners, or how well interventions are designed, adaptation projects have their own underlying political economy and ecology. Adaptation projects can become a flashpoint for competing interests, generating their own sets of winners and losers – even when they might produce a net social gain. Many of these conflicts involve those seeking to enclose resources or exclude stakeholders from access. In some situations, adaptation projects encroach upon and subvert the intended goals of wildlife conservation or the sovereignty of indigenous people. In still other situations, they can entrench disparities in gender, wealth, or social status. In other words, conflict is an unavoidable and inescapable part of the process of adaptation; it can perhaps be managed, but never eliminated. Adaptation and climate policy concern what one study called "a tradeoff between the present and the future, but also a tradeoff between winners and losers at any given time."[2] As such, analysts need to become more rigorous about questioning the underlying assumptions behind a given intervention, more open to the possibility that a project can inequitably distribute benefits, and more accepting that there will *always* be losers.

This conclusion highlights that adaptation is not merely a technical process of infrastructural refinement or climate hardening. Adaptation interventions are about more than technology and economic development; they are about political power, social cohesion, and even ethical and moral concerns over equity, due process, and justice. Brown has written that "climate change requires us to challenge and re-configure ideas about development and to consider alternative strategies... sustainable adaptation is an oxymoron unless it specifically deals with fundamental problems in the dominant paradigm of unsustainable development."[3] In tandem, our analysis suggests that adaptation should be reconceived as a political, deliberative challenge involving the satisfaction of competing preferences; a social dilemma pitting, at times, the climatic and development goals of improved resilience against the pressing needs of marginalized and vulnerable populations; and a moral quandary revolving around how adaptation burdens and benefits are fairly, or unfairly, disseminated.

Drawing from our case studies, in Bangladesh we see local elites and international firms often "win" while poor, landless households with low-wage jobs "lose." In the Netherlands, we see national planners and large construction conglomerates "win" in the erection of a sea wall at the expense of fishers and environmentalists, who "lose." In Hurricane Katrina recovery, large casinos, firms hiring undocumented workers, and corporations with political ties receiving federal aid money "win" while evacuees, racial minorities, and lower and middle-class households "lose." The stalemate within the UNFCCC over the future of adaptation benefits high-income, industrialized countries, who "win" because they are able to perpetuate business as usual while low-income, increasingly vulnerable populations "lose." These examples indicate that adaptation is a double-sided process that can generate, in certain contexts, pain as well as gain.

Our political economy framework also strongly implies that not all agents or agendas of adaptation are benevolent. Some operate similarly to the "roving bandits"[4] that pillage the seas, or the "disaster capitalists"[5] that travel the world seeking to reengineer societies in a state of shock. Some adaptation proponents might in essence act as predators that intentionally seek out and exploit vulnerable communities in order to capitalize on their distress. For these actors, inequitable or suboptimal outcomes are not an accident or unintended byproduct, but a strategic and desirable goal.

This means that stakeholder involvement, while needed to counter exclusion and the dominant interests of some stakeholders, can in some situations be detrimental to the efficacy of an adaptation project. If it is true that some stakeholders actively seek to enclose, exclude, encroach, and entrench, or if their inclusion indirectly contributes to these processes, then their involvement serves to fragment and subvert the intended goals of adaptation. This reality demands that we refocus the discussion about stakeholders away from quantity (more in some cases might not be better), to quality, so that the qualitative goals and interests of stakeholders can be realized. Greater "participation" might not always produce desirable results. Even the noble and intrepid intentions of adaptation planners can dramatically backfire when compromised by political economy processes.

4 Adaptation is interwoven into global processes

Another conclusion is that researchers and practitioners need to reconceive of adaptation as more than a local phenomenon. For instance, a slew of academic research has long suggested that mitigation (stopping emissions) is global in scope, whereas adaptation (coping with consequences) is local in scope. One study tells us that because adaptation efforts are implemented at the local scale, they can enable easier measurement, policy experimentation, democratic community engagement, and new platforms for sharing best-practices between communities.[6] Another expert convincingly argues that the decades-long history of failure to reach a top-down driven international climate change treaty means that actions at smaller scales – local adaptation – might ultimately prove the best chance humanity has at averting disaster.[7] Mitigation is prone to a barrage of common pool resource problems, essentially forcing politicians to sacrifice the present for future gain (which has proven extremely difficult if it does not hold clear promises on return of investments, such as major infrastructure improvements, or strengthened power positions or sense of security, such as taking a nation to war). Conversely, the mantra goes that adaptation projects result in almost immediate local benefits.[8]

Our chapters show, instead, that even seemingly small-scale adaptation interventions at the scale of the community can reinforce broader gender and class distinctions, strengthen the hegemony of markets, create their own type of bureaucracy, or affirm the sovereignty of state institutions. Adaptation's costs and benefits are not necessarily limited to the areas where adaptation projects are being implemented. Projects become intertwined with a global capitalist system seeking to enclose resources and exclude others from access. Even local adaptation efforts often have global effects, with Table 7.2 summarizing the interrelationship among our four case studies, and far-reaching global processes such as capitalism and profiteering, land grabbing, colonialism, and neo-liberalism.

The political economy of adaptation is global in at least two other senses of the term. One is that some of our case studies seem to produce, proliferate, or impose their different political economy features on others, whereas in other situations a project simply consumes or inherits them. In our four cases, we see two originators (for lack of a better term) and two reflectors. The UNFCCC, as an international process, mirrors and then cements the present and historic interests and strategies of dominant negotiating parties or party coalitions. The Dutch Delta Works, because it was a first-of-its-kind project, also became a design template for other low-lying countries or states seeking to erect their own storm surge protective measures – the large, capital intensive nature of the project, the centralized, exclusionary political control – which becomes transplanted around the world as a model for adaptation. Conversely, Bangladesh is largely inheriting the impacts of political economy processes being driven by multinational actors such as the World Bank (providing loans), and transnational companies (offering foreign direct investment for infrastructure or shrimp farms). In the United States, Katrina recovery efforts acted as an accelerant for racial, wage, income, and gender inequality that already existed in stark form across Louisiana and other afflicted communities.

Table 7.2 The global connections to the political economy of adaptation

Case Study	Example(s)
Bangladesh	Hundreds of millions of dollars from international donors are funneled into national infrastructure projects such as polders, embankments, and canals National policies attempt to lure foreign direct investment driving the expansion of industrial shrimp farms A global demand for cheap labor and commodities molds national systems of Bangladeshi governance, and local systems of village hierarchy and patronage focus on profiteering and accumulation of wealth above the welfare of individuals, notably the landless and displaced
Netherlands	The large, centralized, and costly Dutch Delta Works becomes a template for storm surge design exported to other countries including Germany, Italy, the United Kingdom and United States
United States	French and British colonization and slavery lead to centuries of uneven economic development and the creation of class and racial divisions The Gulf of Mexico becomes a major energy producing region and also a port of entry for natural gas and oil, destroying environmental buffers and accelerating industrialization Neo-liberalism and the push for the privatization of government assets and reduced social spending leave the region particularly vulnerable to natural disaster
United Nations	The UNFCCC is an international process enshrined in international law and predicated on a transnational pool of experts, with worldwide negotiations that routinely marginalize representatives from Least Developed Countries

Source: Authors.

Another take-away is that despite differing economic structures and cultures – and four very different case studies occurring in distinct regions, with varying technologies, and over unique time frames – we find enclosure, exclusion, entrenchment, and encroachment present. We see them in an emerging, agrarian, feudal-like society – Bangladesh – at the scale of both national infrastructure projects like canals and local coping strategies such as migration or economic diversification. We see them in a liberal European green state – the Netherlands – as it tries to embark upon an ambitious and grand scheme to protect coastal infrastructure with a series of intricate storm surge barriers and dikes. We see them in an advanced post-industrialized economy – the United

States – seeking to recover from a hurricane through the disbursement of billions of dollars of disaster aid. We see them in an intergovernmental convention – the UNFCCC – cutting across an array of adaptation projects, processes, and approaches being promoted by a multitude of development partners.

Though we have only explored four case studies in this book, some of our preliminary research suggests that the 4E processes operate in other locales. Consider enclosure. In Melbourne, Australia, when planners confronted declining rates of precipitation and a drop in the storage capacity of the city's reservoirs, they decided to build the Wonthaggi desalination plant, which necessitated an enclosure of thirteen sites of land of "significant value" to the Bunurong Aboriginal community.[9] In Honduras, disaster recovery efforts in 1998 after Hurricane Mitch enabled private actors to enclose upon activities formerly in the public sphere. Eight private organizations, mostly transnational NGOs, resettled survivors in the Amarateca Valley and founded new communities; the NGOs then took responsibility for water infrastructure, roads, educational facilities, and housing without government oversight, and with devastating results for some communities that saw higher rates of crime and poverty.[10] In Sweden, safeguarding waterfront areas from flooding, storm surge and other extreme weather impacts have tended to appease the interests of wealthy, politically connected stakeholders at the expense of lower and middle-class homeowners and environmentalists; this safeguarding also makes less funds available for other adaptation projects.[11]

In terms of exclusion, in Norway, coastal zone management efforts facilitated by the Planning and Building Act of 1985 benefitted the most dominant and centralized groups of stakeholders – Norwegian municipalities, fishers, farmers, and tourist entrepreneurs – and largely marginalized less represented or decentralized groups such as reindeer herders, community-based groups, and environmental NGOs.[12] In Kivalina, Alaska, a low-lying community facing displacement from the effects of sea level rise, authority over adaptation projects has excluded local input and been ceded to federal government agencies such as the Army Corps of Engineers and private government contractors.[13] In Mozambique, river basin management – specifically dams to better regulate riverine and lake flow rates impacted by climate change – have primarily served the interests of the urban population in Maputo City, as well as Swaziland and South African commercial enterprises, while increasing the vulnerability of local fishers and subsistence farmers situated in estuarine and coastal ecosystems.[14]

In terms of encroachment, in Tanzania, Marine Protected Areas (MPAs) such as the Mnazi-Bay Ruvuma Estuary Marine Park, intended to bolster the resilience of coral reefs, have instead encroached upon the traditional fishing areas of villagers, leading to dependence on energy-intensive farming.[15] In the Maldives, the past decade has seen the Maldivian government support the idea of "safer islands," the basic goal of which has been to harden or climate-proof infrastructure to the risks of sea level rise, storm surge, and saltwater intrusion.[16,17] These actions, however, have interfered with the ability of the shoreline to act as a natural buffer against storm swells, and also undermined conservation efforts aimed at protecting coral reefs, since seawalls have interrupted the ebb and flow of oceanic nutrient cycles.[18]

Finally, numerous cases of entrenchment exist, spanning Burkina Faso, Ethiopia, Ghana, Kenya, Malawi, Pakistan, South Africa, Sudan, Tanzania, and Zimbabwe. In Burkina Faso, livelihood diversification programs seeking to bolster adaptive capacity have instead seen vulnerable households cope with drought by selling off productive assets such as livestock. This has created windfall profits for "those who have the resources needed to take advantage of distress sales and high prices of agricultural commodities," mostly predatory marketers who are able to buy livestock at low prices from distressed farmers only to resell them at great profit in other areas.[19] In Kenya, one study found that relief aid for natural disasters is dispersed according to "political weight," which benefits urban areas first and peripheral, rural areas last. In some cases, smaller, pastoral communities affected by drought have been forced to abandon their assets and move into permanent aid camps, while elites were able to funnel resources to construct "drought castles" enabling them to consolidate and grow their wealth.[20] In Ethiopia, some adaptation projects in the agricultural and animal husbandry sectors have further concentrated resources among wealthier, larger families.[21] In South Africa and Zimbabwe, diversification of livelihoods as a coping strategy to offset declines in agricultural productivity divided labor and decision-making power away from women in smaller households, facilitated a consolidation of land among larger-scale farmers, and widened differentials in wealth between households, reinforcing "class stratification as higher-income earners redirect portions of their agricultural capital to more lucrative nonagricultural activities."[22] In Ghana, some adaptation projects have been primarily directed at satisfying the desires of men at the expense of a greater workload for women.[23] In Malawi, village leaders decided to implement particular measures – such as flood defenses – only for their own cassava gardens.[24] In Pakistan, some flood recovery efforts have only served to further marginalize rural, agrarian landholders. Recovery interventions have become an "exercise in power" that enabled dominant social classes to "consolidate their position within the rural hierarchy," excluding poorer communities in the process.[25] The end result has been greater levels of debt among the poor as well as loss of control of land and access to water.[26] In southern Sudan, climate change and development interventions served to "reinforce the fundamental patterns of Sudan's (hydro)political economy rather than weaken them" and benefitted Sudan's globalized Islamist elites and foreign investors at the expense of local communities.[27] And in Tanzania, farmers have responded to drought by a process of intensification that requires fertilizer input, meaning it favors wealthier farmers and "usually involves greater specialization and investments which, if successful, will result in greater productivity and increased income levels;"[28] these investments might entrench class relations for decades to come.

The implication, here, with our four case studies and 18 examples briefly summarized in the preceding paragraphs, is that the political economy of adaptation seems to operate seemingly independent of factors such as the type of market economy, mode of government, or form of adaptation.

5 Do not abandon adaptation projects, improve them

Our first three conclusions – that political economy can reduce the effectiveness of adaptation, that winners and losers will always exist, and that adaptation can be influenced by broad global currents alongside more particular local events – might lead some readers to misinterpret our stance on adaptation. We do not believe these three findings mean that all adaptation projects should be cancelled, or even that the four specific projects explored here – Bangladesh's NAPA, the Dutch Delta Works, Hurricane Katrina recovery in the United States, and diplomacy under the UNFCCC – should have been abandoned. We set out to show how the political economy processes of adaptation work, and how they can at times distort or mold adaptation projects and processes to the interests of dominant stakeholders, *not* that they completely undermine or obfuscate all of the benefits of adaptation. Even the specific critiques we raise, some of them quite sobering, are aimed at a target: improving and learning from adaptation's political economy so that the least vulnerable are helped, and so that benefits and burdens are made visible, and distributed fairly and according to representative processes.

We argue that planners and practitioners of adaptation projects need to become more cognizant of the potential for projects to harm others, or admit complicity in the processes of enclosure, exclusion, encroachment, and entrenchment. Adaptation advocates need to become what sociologists call "self-reflexive" by nurturing a habit of skepticism concerning their own knowledge, assumptions, and expectations. Sociologist Steve Woolgar refers to this as benign introspection.[29] In short, adaptation advocates need to continually ask, "Adaptive capacity *for whom*?," "*Who* benefits from this adaptation project?" and "Who can stand to lose, now or in the *future*?"

To give the process of asking such questions a more pragmatic structure, we maintain that a comprehensive set of policy mechanisms are needed to better manage adaptation and to ensure that projects benefit the most vulnerable. As Table 7.3 summarizes, we hold that five policy mechanisms, matched to five corresponding principles, can do much to minimize and manage the political and economic dangers of adaptation projects.

The point of these mechanisms is that they would meet principles of distribute justice and equity, ensuring that benefits and costs from adaptation projects are disseminated fairly, and according to the position of a Rawlsian social contract that most people would accept. They meet principles of due process and procedural justice, ensuring adequate, meaningful representation and engagement in decision-making practices. They also meet principles of open access to information and transparency, demanding that all stakeholders are informed about adaptation projects, that projects are monitored, and that mechanisms exist for the fair exchange of feedback. They meet principles of accountability and responsibility, requiring that adaptation sponsors, investors, planners, and participants are held accountable for the ultimate impacts of compliance, and that strong penalties for noncompliance exist. They meet principles of arbitration and negotiation, ensuring that aggrieved parties have access to remedies and legal recourse when they are wronged.

In addition, the five interconnected policy mechanisms work well at counting the dangers of enclosure, exclusion, encroachment, and entrenchment. Pricing the externalities from adaptation ensures that the often hidden costs of projects become factored in, as best as they can, to market prices, making them more visible to project staff, regulators, and even consumers. Practicing critical stakeholder assessment can jump-start dialogue and facilitate discussions among previously disconnected actors, promote common understanding of key agendas, and incentivize cooperation or identify when views are irresolvable. Free prior informed

Table 7.3 Summary of adaptation design principles and best practices

Principle	Explanation	Corresponding Policy Mechanism	Explanation
Distributive justice and equity	Proportional equivalence of costs and benefits that avoid having some users or adaptation actors get all of the benefits at little cost	Tax adaptation's externalities	Incorporate the full price of adaptation (including as many of its externalities that can be priced) into project budgets, perhaps through an "adaptation tax" similar to a carbon tax, funded through an international mechanism reflecting historical responsibility and macro-economic capacity
Procedural justice and organizational multiplicity	The active involvement of a diverse number of stakeholders, usually brought about by participatory modes of decision-making or incorporation of local actors in the planning process, involving multiple institutions to deepen coverage and offer parallel systems of governance	Critically involve stakeholders	Perform and practice critical stakeholder analysis to understand interests and power relationships of allies and competitors whose involvement can impact the success or failure of averting the externalities of an adaptation project.
Information and transparent monitoring	The distribution of data about costs and benefits, the close monitoring of programs, and mechanisms that encourage feedback and the sharing of knowledge	Adhere to free prior informed consent	Maintain a robust consultative process whereby a potentially affected community engages in an open and informed dialogue with individuals or other persons interested in pursuing activities in the area or areas occupied or traditionally used by the affected community
Accountability and responsibility	Project sponsors bear some of the costs of governance themselves and are accountable for their actions and also that graduated sanctions exist to enforce rules and penalize noncompliance	Implement punitive bonds and inspection panels	Environmental bonds would require adaptation sponsors to purchase a sort of "insurance" that would be voided or forfeited when projects damage communities or the environment, truth commissions and inspection panels would be there to collect facts and report on potential grievances
Adaptability and arbitration	Mechanisms are in place for conflict resolution and to respond to challenges and unforeseen events	Create an international tribunal for redress	Establishing a Tribunal for Adaptation Justice and Conflict Resolution would provide the opportunity to handle grievances and to give redress in the form of restitution, compensation, rehabilitation, satisfaction, and guarantees of non-repetition

Source: Authors.

consent reduces the risk that no coercion, intimidation, or manipulation has occurred in the design or implementation of an adaptation project; that consent has been sought sufficiently in advance of any meaningful decision to proceed; that information about the project is provided that covers its nature, size, pace, reversibility, and scope; and that agreement with proposed project measures exists. Environmental bonds for adaptation would provide a contractual guarantee that the principal of the bond (an adaptation project manager, for instance) performs in an environmentally benign, and socially responsible, manner; bonds would be levied for the current best estimate of the largest potential damages. Adaptation truth commissions or inspection panels offer two useful tools for documenting potential abuses and grievances. An international Tribunal for Adaptation Justice and Conflict Resolution could then give redress and reparation to those actually harmed.

In sum, pursuing these five integrated principles and policies enables us to refine and enhance the process of climate change adaptation. We want our framework to politicize and expose the pathologies of adaptation, we do not want readers concluding that adaptation ought to be minimized in international climate change efforts. On the contrary. The medicine may have some unfortunate side effects, but it still needs to be taken.

6 Further research is needed

Our final conclusion is that more research is needed to refine our framework, strengthen it, or even contest it. For instance, here we have tested only the dependent variable. Research could explore whether the inverse is true, or whether situations exist where enclosure, exclusion, encroachment, and entrenchment have somehow benefitted communities, or how communities can utilize various mechanisms to break from their systemic nature. How unfair distribution of adaptation comes about could be within the purview of a review paper, along with assessing the interrelationship among the four processes. Are some of them more damaging than others, or do they deserve to be weighted equally? What hypotheses does our political economy typology give rise to, and how can these be rigorously tested? Do some, many, or even most adaptation projects suffer from these political economy processes, or only a few? Currently, it is difficult to determine whether the political economy of adaptation is pathological, affecting almost all projects, or a rarity.

Other questions relate to what might be unique about adaptation's political economy compared to, say, the political economy of public health or national security. Given how close adaptation projects often sit to ordinary economic development projects – in terms of the actors involved, and many of the broader targets set – are the political economy factors we identify truly unique to adaptation, or are they involved in many types of ordinary development projects, from building bridges to bolstering food security? Put another way, are we merely depicting the negative impacts of capitalism, hegemony, and greed – factors hardly isolated to adaptation projects? As of now, we have little sense of how adaptation relates in comparison to the political economy processes (our 4Es, or other ones entirely) afflicting other technological and industrial sectors.

Lastly, though some of the analysis in this book draws from earlier fieldwork of the authors in Bangladesh (Sovacool) and at UNFCCC COP delegations (Linnér), most of it was done using a literature review of secondary sources. Thus, proving or disproving our theoretical framework with original data collection in the form of field research, household interviews, surveys, focus groups, participant observation, and the like would be most fruitful.

That said, despite these promising areas of future inquiry and potential gaps in our own analysis, we still know enough about the political economy of adaptation to conclude that researchers need to take a more critical view of projects, and seek to better comprehend their deeper implications. Done well, adaptation can build resilience and enable humanity to cope with climate change. Done poorly, however, it can enable the elite capture of resources, negatively affect land tenure, exclude indigenous people, marginalize stakeholders, transfer wealth from public to private hands, or aggravate social inequality. Adaptation can, perhaps ironically, even require further environmental adaptation – responding to the beleaguered and negative impacts of an earlier adaptation effort. This demands, as adaptation researchers Richard Klein and Sirkku Juhola have argued,[30] that we reframe our definition of adaptation from "responding to the impacts of climate change" to "responding to the impacts of climate change as well as the earlier impacts of climate policy, including previous adaptation projects." Our four case studies support this contention as well, given the landlessness and poverty impacts of some adaptation processes in Bangladesh, the fact that the largest piece of the Dutch Delta Works (the Eastern Scheldt Storm Surge Barrier) may need to be entirely removed and expensively decommissioned if sea levels rise beyond one meter, the unequal levee protection of Katrina reconstruction, and the lock-in toward technological inferiority being facilitated by the UNFCCC.

Thus, we in the climate policy community need to move well beyond the type of thinking that "a lost opportunity can be as bad as a negative impact."[31] In some cases, adaptation *is* the negative impact, and given the resources and transaction costs involved in implementing projects, it would have been better if nothing had been done in the first place. For perhaps this reason alone, we need to rethink our fundamental assumptions about when adaptation projects are warranted, and the criteria by which we ought to evaluate their success.

Notes

1 Introduction to the Political Economy of Climate Change Adaptation

1. McCarthy, J. (2005). Commons as counterhegemonic projects. *Capitalism, Nature, Socialism*, 16(1), 9–24.
2. Kondro, W. (2004). Canada: Monsanto wins split decision in patent fight over GM crop. (May 28). *Science*, 304(5675), 1229.
3. Ford, J.D., Berrang-Ford, L., & Paterson, J. (2011). A systematic review of observed climate change adaptation in developed nations. *Climatic Change*, 106, 327–336.
4. Bryan, E. & Behrman, J. (2013). Community-based adaptation to climate change: A theoretical framework, overview of key issues and discussion of gender differentiated priorities and participation. CAPRi Working Paper No. 109. Washington, DC: International Food Policy Research Institute, p. 12.
5. Biermann, F., Pattberg, P., & Zelli, F. (2010). Global climate governance beyond 2012: Architecture, agency, and adaptation. In M. Hulme & H. Neufeldt (eds), *Making Climate Change Work For Us: European Perspectives on Adaptation and Mitigation Strategies* (pp. 263–290). Cambridge, UK: Cambridge University Press.
6. Adger, W.N. (2006). Vulnerability. *Global Environmental Change*, 16, 268–281.
7. Sweeney, G., Dobson, R., Despota, K., & Zinnbauer, D. (eds) (2011). *Global Corruption Report: Climate Change*. London, UK: Earthscan.
8. Michaelowa, A. (2000). The relative strength of economic interests in shaping EU climate policy: A hypothesis. *Energy and Environment*, 11(3), 277–292.
9. Schreurs, M.A. & Tiberghien, Y. (2007). Multi-level reinforcement: Explaining European Union leadership in climate change mitigation. *Global Environmental Politics*, 7(4), 19–46.
10. Barnett, J. & O'Neil, S. (2010). Maladaptation. *Global Environmental Change*, 20(2), 211–213.
11. IPCC. (2012). Managing the risks of extreme events and disasters to advance climate change adaptation. A Special Report of Working Groups I and II of the Intergovernmental Panel on Climate Change. 1pp. Environmental Change, volthat hides data/maps from other Yale userson product (PolicyMap). I am on reference, please comeCambridge, UK: Cambridge University Press.
12. Ruhl, J.B. (2012). The political economy of climate change winners. *Minnesota Law Review*, 97, 207–277.
13. Füssel, H.-M. (2010). How inequitable is the global distribution of responsibility, capability, and vulnerability to climate change? A comprehensive indicator-based assessment. *Global Environmental Change*, 20(4), 597–611.
14. Eriksen, S., et al. (2011). When not every response to climate change is a good one: Identifying principles for sustainable adaptation. *Climate and Development*, 3(1), 7–20.

15. Wilbanks, T.J., et al. (2003). Integrating mitigation and adaptation as possible responses to global climate change. *Environment*, 45(5), 28–38.
16. Tol, R.S.J. (2005). Adaptation and mitigation: Trade-offs in substance and methods. *Environmental Science & Policy*, 8(6), 572–578.
17. Wilbanks, T.J. (2005). Issues in developing a capacity for integrated analysis of mitigation and adaptation. *Environmental Science & Policy*, 8(6), 541–547.
18. Klein, R.J.T., Schipper, E.L.F., & Dessai, S. (2005). Integrating mitigation and adaptation into climate and development policy: Three research questions. *Environmental Science & Policy*, 8(6), 579–588.
19. Bankoff, G. (2001). Rendering the world unsafe: "Vulnerability" as a western discourse. *Disasters*, (1), 19–35.
20. Adger, W.N., Benjaminsen, T.A., Brown, K., & Svarstad, H. (2001). Advancing a political ecology of global environmental discourses. *Development and Change*, 32(4), 681–715.
21. Taylor, M. (2014). *The Political Ecology of Climate Change Adaptation: Livelihoods, Agrarian Change, and the Conflicts of Development* (p. 3). London, UK: Routledge.
22. Sovacool, B.K., Linner, B.-O. & Goodsite, M.E. (2015). The political economy of climate adaptation. *Nature Climate Change* 5(7): p. 616–618.
23. Intergovernmental Panel on Climate Change (IPCC), M.L. Parry, O.F. Canziani, J.P. Palutikof, P.J. Van der Linden, & C.E. Hanson (eds) (2007). Climate Change 2007: Impacts, adaptation, and vulnerability. Contribution of Working Group II to the Fourth Assessment Report of the Intergovernmental Panel on Climate Change (p. 9). Cambridge, UK: Cambridge University Press.
24. IPCC (2007).
25. Brown, M.A. & Sovacool, B.K. (2011). *Climate Change and Global Energy Security: Technology and Policy Options*. Cambridge, MA: MIT Press.
26. The Economics of Climate Adaptation Working Group (2009). *Shaping Climate Resilient Development*. New York, NY: ClimateWorks Foundation, Global Environment Facility, European Commission, McKinsey & Company, The Rockefeller Foundation, Standard Chartered Bank and Swiss Re.
27. Ahmed, A.U., Alam, M., & Rahman, A.A. (2011). Adaptation to climate change in Bangladesh: Future outlook. In S. Huq, Z. Karim, M. Asaduzzaman, & F. Mahtab (eds), *Vulnerability and Adaptation to Climate Change for Bangladesh* (pp. 125–143). London, UK: Kluwer Academic Publishers.
28. Quoted in Romm, J. (2013). White House: "We are going to achieve the President's goal of reducing greenhouse gas emissions by 17 percent by 2020." *Climate Progress*. January 22. Available at: http://thinkprogress.org/climate/2013/01/22/1478741/white-house-achieve-presidents-goal-of-reducing-greenhouse-gas-emissions-by-17-percent-by-2020/.
29. Klein, R.J.T. (2011). Adaptation to climate change: More than technology. In I. Linkov & T.S. Bridges (eds), *Climate: Global Change and Local Adaptation* (pp. 159–170). London, UK: Springer.
30. IPCC (2012).
31. IPCC (2014). Climate change 2014: Synthesis report. Contribution of Working Groups I, II and III to the Fifth Assessment Report of the Intergovernmental Panel on Climate Change. Core Writing Team, R.K. Pachauri & L.A. Meyer (eds). Geneva, Switzerland: IPCC.

32. Kelman, I. & Gaillard, J.C. (2010). Embedding climate change adaptation within disaster risk reduction. *Environment and Disaster Risk Management*, 4, 23–46.
33. Ensor, J. & Berger, R. (2009). *Understanding Climate Change Adaptation: Lessons From Community-based Approaches*. New York, NY: Practical Action Publishing.
34. Adger (2006).
35. O'Brien, K., Eriksen, S., Nygaard, L.P., & Schjolden, A. (2007). Why different interpretations of vulnerability matter in climate change discourses. *Climate Policy*, 7(1), 73–88.
36. O'Brien et al. (2007).
37. O'Brien et al. (2007).
38. O'Brien et al. (2007).
39. Barnett & O'Neill (2010); Bisaro et al. (2010).
40. McDaniels, T., Chang, S., Cole, D., Mikawoz, J., & Longstaff, H. (2008). Fostering resilience to extreme events within infrastructure systems: Characterizing decision contexts for mitigation and adaptation. *Global Environmental Change*, 18(2), 310–318.
41. Marshall, N.A. (2010). Understanding social resilience to climate variability in primary enterprises and industries. *Global Environmental Change*, 20, 36–43.
42. Venema, H.D. & Cisse, M. (2004). Seeing the light: Adapting to climate change with decentralized renewable energy in developing countries. (Toronto, Canada: International Institute for Sustainable Development), p. 11.
43. Folke, C. (2006). Resilience: The emergence of a perspective for social-ecological systems analyses. *Global Environmental Change*, 16(3), 253–267.
44. Ensor & Berger (2009).
45. Feenstra, J.F., Burton, I., Smith, J.B., & Tol, R.S.J. (eds). (1998). *Handbook on Methods for Climate Change Impact Assessment and Adaptation Strategies*. Amsterdam, the Netherlands: Institute for Environmental Studies, Vrije University, Amsterdam; UNEP Headquarters, Atmosphere Unit, p. 5–4..
46. IPCC (2007) p. 378.
47. Edwards, P.N. (2010). *A Vast Machine: Computer Models, Climate Data, and the Politics of Global Warming* (p. 7). Cambridge, MA: MIT Press.
48. Archer, D. & Rahmstorf, S. (2010). *The Climate Crisis: An Introductory Guide to Climate Change*. Cambridge, MA: Cambridge University Press.
49. Hay, C.C., et al., (2015). Probabilistic reanalysis of twentieth-century sea-level rise. *Nature*, 517(7535), 481–484.
50. Schneider, S.H. (2010). Geo-engineering: Could we or should we make it work? In B. Launder & J.M.T. Thompson (eds), *Geo-engineering Climate Change: Environmental Necessity or Pandora's Box?* (pp. 3–26, 23). Cambridge, UK: Cambridge University Press.
51. Lenton, T.M., et al. (2008). Tipping elements in the earth's climate system. *Proceedings of the National Academy of Sciences*, 105, 1786–1793.
52. Thompson, J.M.T. & Sieber, J. (2010). Predicting climate tipping points. In B. Launder & J.M.T. Thompson (eds), *Geo-engineering Climate Change: Environmental Necessity or Pandora's Box?* (pp. 50–83). Cambridge, UK: Cambridge University Press.
53. See Lenton, et al. (2008); Shakhova, N., Semiletov, I., Salyuk, A., Yusupov, V., Kosmach, D., Gustafsson, O. (2010). Extensive methane venting to the

atmosphere from sediments of the East Siberian Arctic Shelf. *Science*, 327, 1246–1250.
54. *The Economist*. (2012, June 16). The melting north [Special Report on The Arctic], 3–4.
55. *The Economist*, 2012.
56. Mann, M.E. (2009). Defining dangerous anthropogenic interference. *PNAS*, 106(11), 4065–4066.
57. Kousky, C. & Cooke, R.M. (2009). The unholy trinity: Fat tails, tail dependence, and micro-correlations. Resources for the Future Discussion Paper No. 09–36-REV.
58. Archer & Rahmstorf (2010).
59. Barrett, S. (2009). Climate treaties and the imperative of enforcement. In D. Helm & C. Hepburn (eds), *The Economics and Politics of Climate Change* (pp. 58–80). Oxford, UK: Oxford University Press.
60. Aaheim, A., et al. (2010). National responsibilities for adaptation strategies: Lessons from four modeling frameworks. In M. Hulme & H. Neufeldt (eds), *Making Climate Change Work For Us: European Perspectives on Adaptation and Mitigation Strategies* (pp. 87–112, 88). Cambridge, UK: Cambridge University Press.
61. *The Economist* (2010). Adapting to climate change. November 27, pp. 79–82.
62. Klein, R.J.T., et al. (2007). Inter-relationships between adaptation and mitigation. In M.L. Parry, O.F. Canziani, J.P. Palutikof, P.J. Van der Linden, & C.E. Hanson (eds), Climate change 2007: Impacts, adaptation, and vulnerability. Contribution of Working Group II to the Fourth Assessment Report of the Intergovernmental Panel on Climate Change (pp. 745–777). Cambridge, UK: Cambridge University Press.
63. Sovacool, B.K. & Brown, M.A. (2009). Scaling the response to climate change. *Policy & Society*, 27(4), 317–328.
64. Moser, S.C. (2012). Adaptation, mitigation, and their disharmonious discontents: An essay. *Climatic Change*, 111, 165–175.
65. Moser (2012).
66. Moser (2012).
67. Sovacool, B.K. (2013). *Energy and Ethics* (pp. 181–182). New York, NY: Palgrave Macmillan.
68. The Economics of Climate Adaptation Working Group (2009).
69. Tran, P., Tran, T.H., & Tran, A.T. (2014). *Sheltering From a Gathering Storm: Typhoon Resilience in Vietnam*. Da Nang, Vietnam: Institute for Social and Environmental Transition-International.
70. Asian Development Bank (2009). *The Economics of Climate Change in Southeast Asia: A Regional Review* (April). Manila, Philippines: ADB.
71. Biermann, Pattberg, & Zelli (2010) p. 280.
72. Prouty, A.E. (2009). The clean development mechanisms and its implications for climate justice. *Columbia Journal of Environmental Law*, 34(2), 513–540.
73. Hope, C. (2009). How deep should the deep cuts be? Optimal CO_2 emissions over time under uncertainty. *Climate Policy*, 9(1), 3–8; De Brun, K., Dellink, R., & Agrawala, S. (2009). *Economic Aspects of Adaptation to Climate Change: Integrated Assessment Modeling of Adaptation Costs and Benefits*. Paris, France: OECD.
74. Gupta, J., Persson, A., & Olsson, L. (2010). Mainstreaming climate change in development cooperation policy: Conditions for success. In M. Hulme & H.

Neufeldt (eds), *Making Climate Change Work For Us: European Perspectives on Adaptation and Mitigation Strategies* (pp. 319–339). Cambridge, UK: Cambridge University Press.

75. Peake, S. & Smith, J. (2009). *Climate Change: From Science to Sustainability* (p. 135). Oxford, UK: Oxford University Press.
76. IPCC (2012).
77. Smith, K.R., Desai, M.A., Rogers, J.V., & Houghton, R.A. (2013). Joint CO2 and CH4 accountability for global warming, July 11. *Proceedings of the National Academy of Sciences: Early Edition.*
78. Quoted in *The Economist* (2008). Climate change and the poor: Adapt or die. September 11.
79. Quoted in Belt, D. (2011). Buoyant Bangladesh and the coming storm. *National Geographic*, 219(5), 58–83.
80. Roger Jr. Pielke (2014). *The Rightful Place of Science: Disasters and Climate Change*. Tempe, AZ: Consortium for Science, Policy & Outcomes, 2014).
81. Intergovernmental Panel on Climate Change and C.B. Field (2012). Managing the risks of extreme events and disasters to advance climate change adaptation: Special report of the Intergovernmental Panel on Climate Change (p. 268). Cambridge: Cambridge University Press.
82. Field (2012) p. 268.
83. International Council on Human Rights Policy (2008). *Climate Change and Human Rights: A Rough Guide*. Versoix Switzerland: International Council on Human Rights Policy.
84. Gardiner, S.M. (2010). Ethics and global climate change. In S.M. Gardiner, S. Caney, D. Jamieson, & H. Shue (eds), *Climate Ethics: Essential Readings* (pp. 3–35). Oxford, UK: Oxford University Press,.
85. Caney, S. (2005). Cosmopolitan justice, responsibility, and global climate change. *Leiden Journal of International Law*, 18, 747–775. See also, Neumayer, E. (2000). In defence of historical accountability for greenhouse gas emissions. *Ecological Economics*, 33, 185–192.
86. Eriksen, S., et al. (2011). When not every response to climate change is a good one: Identifying principles for sustainable adaptation. *Climate and Development*, 3(1), 7–20.
87. Gilpin, R. (1987). *The Political Economy of International Relations*. Princeton, NJ: Princeton University Press.
88. Caporaso, J.A. & Levine, D.P. (1992). *Theories of Political Economy*. Cambridge, UK: Cambridge University Press.
89. Castree, N. (2010). Neoliberalism and the biophysical environment 2: Theorising the neoliberalisation of nature. *Geography Compass*, 4(12), 1734–1746.
90. Bohle, H.G., Downing, T.E. & Watts, M.J. (1994) Climate change and social vulnerability: Towards a sociology and geography of food insecurity. *Global Environmental Change*, 4(1), 37–48.
91. Wolff, R. & Resnick, S. (1987). *Economics: Marxian Versus Neoclassical*. Baltimore, MD: The Johns Hopkins University Press.
92. Wolf, E. (1972). Ownership and political ecology. *Anthropology Quarterly*, 45(3), 201–205.
93. Biersack, A. & Greenberg, J.B. (2006). *Reimagining Political Ecology*. Durham, NC: Duke University Press.

94. Watts, M.J. (2000). Political ecology. In T. Barnes & E. Sheppard (eds), *A Companion to Economic Geography* (pp. 257–275). Oxford, UK: Blackwell.
95. Bryant, R.L. & Bailey, S. (1997). *Third World Political Ecology* (pp. 28–29). London, UK: Routledge,.
96. Robbins, P. (2004). *Political Ecology: A Critical Introduction*. London, UK: Earthscan, 20.
97. Low, N. & Gleeson, B. (1998). *Justice, Society, and Nature: An Exploration of Political Ecology*. London, UK: Routledge.
98. Schlosberg, D. (1999). *Environmental Justice and the New Pluralism: The Challenge of Difference for Environmentalism*. Oxford, UK: Oxford University Press.
99. Byrne, J. (ed.) (2002). *Environmental Justice: Discourses in the International Political Economy*. New Brunswick, NJ: Transaction Publishers.
100. Bowen, W.M. & Wells, M.V. (2002). The politics and reality of environmental justice. *Public Administration Review*, 62(6), 688–698.
101. Walker, G. (2012). *Environmental Justice: Concepts, Evidence, and Politics*. London, UK: Routledge.
102. Arnold, D.G. (ed.) (2011). *The Ethics of Global Climate Change*. Cambridge, UK: Cambridge University Press.
103. Sovacool, B.K. & Dworkin, M.H. (2014). *Global Energy Justice: Problems, Principles, and Practices*. Cambridge: Cambridge University Press.
104. Sovacool, B.K., Sidortsov, R., & Jones, B. (2013). *Energy Security, Equality, and Justice*. London, UK: Routledge.
105. Sovacool, B.K. (2013). *Energy & Ethics: Justice and the Global Energy Challenge*. Basingstoke, UK/New York, USA: Palgrave.
106. Sovacool, B.K. (2013). Adaptation: The complexity of climate justice. *Nature Climate Change*, 3, 959–960.
107. Escobar, A. (1995). *Encountering Development: The Making and Unmaking of the Third World*. Princeton, NJ: Princeton University Press.
108. Scott, J.C. (1998). *Seeing Like a State: How Certain Schemes to Improve the Human Condition Have Failed*. New Haven, CT: Yale University Press.
109. Nelson, N. & Wright, S. (1995). *Power and Participatory Development: Theory and Practice*. London, UK: Intermediate Technology Publications.
110. Schneider, H. & Libercier, M.-H. (eds) (1995). *Participatory Development: From Advocacy to Action*. Paris: OECD.
111. Pieterse, J. (2001). *Development Theory: Deconstructions/Reconstructions*. London: SAGE.
112. Ferguson, J.A. (1994). *The Anti-Politics Machine: Development, Depoliticization, and Bureaucratic Power in Lesotho*. Minneapolis, MN: University of Minnesota Press.
113. Castree (2010) p. 1739.
114. Faber, D. & McCarthy, D. (2003). Neo-liberalism, globalization, and the struggle for ecological democracy: Linking sustainability and environmental justice. In J. Agyeman, et al. (eds), *Just Sustainabilities: Development in an Unequal World*. Cambridge, MA: MIT Press; Watts, M. & Peet, R. (2004). Liberating political ecology. In R. Peet & M. Watts (eds), *Liberation Ecologies: Environment, Development, and Social Movements* (pp. 3–47). London, UK: Routledge.
115. Forsyth, T. (2008). Political ecology and the epistemology of social justice. *Geoforum*, 39, 756–764.

116. Ryan, S. (2014). Rethinking gender and identity in energy studies. *Energy Research & Social Science*, 1(1), 96–105.
117. McCarthy, J. (2009). Commons. In N. Castree, et al. (eds), *A Companion to Environmental Geography* (pp. 498–514). Oxford, UK: Wiley-Blackwell.
118. Hoogvelt, A. (1987). *Globalization and the Postcolonial World: The New Political Economy of Development* (p. 3). Baltimore, MD: Johns Hopkins University Press.
119. Prudham, W.S. (2009). Commodification. In N. Castree, et al. (eds), *A Companion to Environmental Geography* (pp. 123–142). Oxford, UK: Wiley-Blackwell.
120. Harvey, D. (2004). The "new" imperialism: Accumulation by dispossession. *Socialist Register*, 40, 63–87.
121. Harvey, D. (2003). *The New Imperialism* (p. 148). Oxford, UK: Oxford University Press.
122. Bridge, G. (2008). Global production networks and the extractive sector: Governing resource-based development. *Journal of Economic Geography*, 8(3), 389–419.
123. White, B., Borras, S.M., Jr., Hall, R., Scoones, I., & Wolford, W. (2012). The new enclosures: Critical perspectives on corporate land deals. *The Journal of Peasant Studies*, 39(3–4), 619–647.
124. Hamza, M. & Corendea, C. (eds) (2012). Climate change and fragile states: Rethinking. *Studies of the University Research, Counsel, Education Publication Series of UNU-EHS No. 16/2012*. Bonn, Germany: United Nations University Institute for Environment and Human Security.
125. Ghani, A., Lockhart, C., & Carnahan, M. (2005). Closing the sovereignty gap: An approach to state-building. Working Paper, No. 253. London: Overseas Development Institute (ODI), p. 19.
126. Alaniz, R. (2012). Unsupervised recovery: Adaptation strategies by two NGOs in post-Mitch Honduras. In M. Hamza & C. Corendea (eds), *Climate Change and Fragile States: Rethinking*. Studies of the University Research, Counsel, Education Publication Series of UNU-EHS No. 16/2012 (pp. 112–130). Bonn, Germany: United Nations University Institute for Environment and Human Security.
127. Poocharoen, O.-O. & Sovacool, B.K. (2012). Exploring the challenges of energy and resources network governance. *Energy Policy*, 42, 409–418.
128. Sovacool, B.K. & Florini, A.E. (2012). Examining the complications of climate and energy governance. *Journal of Energy and Natural Resources Law*, 30(3), 235–263.
129. McFarlane, A. (2007–2008). Fighting for the high ground: Race, class, markets and the development done right in post Katrina recovery. *Washington & Lee Journal of Civil Rights & Social Justice*, 14, 77–96.
130. Heynen, N. & Robbins, P. (2007). The neoliberalization of nature: Governance, privatization, enclosure and valuation. *Capitalism Nature Socialism*, 16(1), 5–8.
131. Robbins, P. (2012). *Political Ecology: A Critical Introduction* (pp. 22–23). London, UK: Wiley & Sons.
132. Satterthwaite, D., Dodman, D., & Bicknell, J. (2009). Conclusions: Local development and adaptation. In J. Bicknell, D. Dodman & D. Satterthwaite

(eds), *Adapting Cities to Climate Change Understanding and Addressing the Development Challenges* (pp. 359–383). London, UK: Earthscan.

133. Few, R., Brown, K., & Tompkins, E.L. (2007). Public participation and climate change adaptation: Avoiding the illusion of inclusion. *Climate Policy*, 7(1), 46–59.
134. Marini, R.M. (2005). Dialética da dependência. In R. Traspadini & J.P. Stedile (eds), *Ruy Mauro Marini: Vida e obra* (pp. 137–180). São Paulo, Brazil: Expressão Popular.
135. Adger, W.N., Paavola, J., & Huq, S. (2006). Toward justice in adaptation to climate change. In W.N. Adger, J. Paavola, S. Huq, & M.J. Mace (eds), *Fairness in Adaptation to Climate Change* (pp. 1–19). Cambridge, MA: MIT Press.
136. Gilpin (1987) pp. 36–38.
137. Harvey (2004).
138. See Harvey (2003); Harvey, D. (2006). *Spaces of Global Capitalism: Towards a Theory of Uneven Geographical Development*. London, UK: Verso.
139. Cooke, B. & Kothari, U. (2001). *Participation: The New Tyranny?* New York, NY: Zed Books.
140. For a sample, see Brockington, D. & Igoe, J. (2006). Eviction for conservation: A global overview. *Conservation and Society*, 4(3), 424–470; Igoe, J., Neves, K., & Brockington, D. (2010). A spectacular eco-tour around the historic bloc: Theorising the convergence of biodiversity conservation and capitalist expansion. *Antipode*, 42(3), 486–512; Peluso, N.L. & Lund, C. (2011). New frontiers of land control: Introduction. *Journal of Peasant Studies*, 38(4), 667–681; Bryant, R.L. (1998). Power, knowledge and political ecology in the third world: a review. *Progress in Physical Geography*, 22: 79–94.
141. Turner, W.R., Bradley, B.A., Estes, L.D., Hole, D.G., Oppenheimer, M. & Wilcove, D.S. (2010). Climate change: Helping nature survive the human response. *Conservation Letters*, 3(5), 304–312.
142. Klein et al. (2007). (pp. 745–777).
143. Sovacool, B.K. & Brown, M.A. (2007). Scaling the policy response to climate change. *Policy and Society*, 27, 317–328.
144. For more on these types of tradeoffs, see Klein, Schipper, & Dessai (2005); Wilbanks et al., 2003; Wilbanks (2005).
145. Moser (2012).
146. Moser (2012).
147. Moser (2012).
148. Castree, N. (2003). Commodifying what nature? *Progress in Human Geography*, 27(3), 273–297; Castree (2010).
149. Sovacool, B.K. (2011). Conceptualizing hard and soft paths for climate change adaptation. *Climate Policy*, 11(4), 1177–1183.
150. Biermann, Pattberg, & Zelli (2010).
151. Little, P.D., et al. (2001). Avoiding disaster: Diversification and risk management among East African Herders. *Development and Change*, 32(3), 401–433.
152. Holling, C.S., Gunderson, L.H. & Ludwig, D. (2002). In quest of a theory of adaptive change. In L.H. Gunderson & C.S. Holling (eds), *Panarchy:*

Understanding Transformations in Human and Natural Systems (pp. 3–22). Washington DC: Island Press.

153. Christens, B. & Speer, P.W. (1996). Tyranny/transformation: Power and paradox in participatory development. *Qualitative Social Research*, 7(2), 1–10.
154. Ruhl (2012).
155. Dutta, D. (2009). *Elite Capture and Corruption: Concepts and Definitions* (October). New Delhi, India: National Council of Applied Economic Research.
156. Mirza, M.Q. (2009). Climate change and adaptation: Accounting for flood hazard in Bangladesh. In R.D. van den Berg & O. Feinstein (eds), *Evaluating Climate Change and Development* (pp. 299–318). New Brunswick, NJ: Transaction Publishers.
157. Bullard, R.D. & Wright, B. (2009). Introduction. In R.D. Bullard & B. Wright (eds), *Race, Place, and Environmental Justice After Hurricane Katrina: Struggles to Reclaim, Rebuild, and Revitalize New Orleans and the Gulf Coast* (pp. 1–13, 1). Philadelphia, PA: Westview Press.
158. Stern, P.C. (2014,). Energy: We need all hands on deck. *Nature*, 513(September 4), 33.
159. Merton, R. (1969). Foreword. In A.H. Barton, *Communities in Disaster: A Sociological Analysis of Collective Stress Situations* (pp. vii–xxxvii, xi). Garden City, NY: Doubleday.
160. George, A.L. & Bennett, A. (2004). *Case Studies and Theory Development in the Social Sciences* (p. 5). Cambridge, MA: Harvard University Press.
161. Flyvbjerg, B. (2006). Five misunderstandings about case study research. *Qualitative Inquiry*, 12(2), 219–245; Flyvbjerg, B. (2001). *Making Social Science Matter: Why Social Inquiry Fails And How It Can Succeed Again*. Cambridge, MA: Cambridge University Press.
162. Yin, R.K. (2003). *Case Study Research: Design and Methods*. London, UK: SAGE.
163. Baxter, P. & Jack, S. (2008). Qualitative case study methodology: Study design and implementation for novice researchers. *The Qualitative Report*, 13(4), 544–559.
164. Flyvberg (2001, 2006).
165. Lijphart, A. (1971). Comparative politics and the comparative method. *American Political Science Review*, 65(3), 682–693.
166. Glaser, B.G. & Strauss A.L. (1997). *The Discovery of Grounded Theory: Strategies for Qualitative Research*. New York, NY: Aldine de Gruyter.
167. King, G., Keohane, R.O. & Verba, S. (1994). *Designing Social Inquiry: Scientific Inference in Qualitative Research*. Princeton, NJ: Princeton University Press.
168. George & Bennett (2004).
169. McCarthy (2005).
170. Bebbington, A. & Bury J. (eds). (2013). *Subterranean Struggles: New Dynamics of Mining, Oil, and Gas in Latin America*. Austin, TX: University of Texas Press.
171. Böhm, S., Misoczky, M.C., & Moog, S. (2012). Greening capitalism? A Marxist critique of carbon markets. *Organization Studies*, 33(11), 1617–1638.

2 Bamboo Thumping Bandits: The Political Economy of Climate Adaptation in Bangladesh

1. Richard, M.A. (2007). Climate change and human security. In J.F.C. DiMento & P. Doughman (eds), *Climate Change: What It Means For Us, Our Children, and Our Grandchildren* (pp. 161–180). Cambridge, MA: MIT Press.
2. Mirza, M.M.Q. (2002). Global warming and changes in the probability of occurrence of floods in Bangladesh and implications. *Global Environmental Change*, 12(2), 127–138.
3. Rawlani, A. & Sovacool, B.K. (2011). Building responsiveness to climate change through community based adaptation in Bangladesh. *Mitigation and Adaptation Strategies for Global Change*, 16(8), 845–863.
4. Rawlani & Sovacool (2011).
5. Ahmed, A.U. (2006). *Bangladesh Climate Change Impacts and Vulnerability: A Synthesis* (July). Dhaka, Bangladesh: Department of Environment, Climate Change Cell.
6. Rawlani & Sovacool (2011).
7. Ali, A. (1996). Vulnerability of Bangladesh to climate change and sea level rise through tropical cyclones and storm surges. *Journal of Water Air Soil Pollution*, 92, 171–179; Ali, A. (1999). Climate change impacts and adaptation assessment in Bangladesh. *Climate Research*, 12, 109–116; Huq, S. & Asaduzzaman, M. (2010). Overview. In S. Huq, Z. Karim, M. Asaduzzaman, & F. Mahtab (eds), *Vulnerability and Adaptation to Climate Change for Bangladesh* (pp. 1–11). London, UK: Kluwer Academic Publishers; Sajjaduzzaman, M., Muhammed, N., & Koike, M. (2005). Mangrove plantation destruction in Noakhali coastal forests of Bangladesh: A case study on causes, consequences and model prescription to halt deforestation. *International Journal of Agriculture and Biology*, 7(5), 732–734.
8. Chowdhury, N.J. (2008). A journey towards development: The impact of local NGO programmes on women living in the char lands of Bangladesh. *Development in Practice*, 18(1), 117–124.
9. Chowdhury (2008).
10. Chowdhury (2008).
11. Ahammad, R., Nandy, P., & Husnain, P. (2013). Unlocking ecosystem based adaptation opportunities in coastal Bangladesh. *Journal of Coastal Conservation*, 17(4), 833–840.
12. Rawlani & Sovacool (2011).
13. Belt, D. (2011). Buoyant Bangladesh and the coming storm. *National Geographic*, 219(5), 58–83.
14. Belt (2011).
15. Mirza (2002).
16. Government of Bangladesh (2005). *National Adaptation Program of Action* (NAPA). Dhaka, Bangladesh: Ministry of Environment and Forests (MOEF).
17. Agrawala, S., Ota, T., Ahmed, A.U., Smith, J., & van Aalst, M. (2003). *Development and Climate Change in Bangladesh: Focus on Coastal Flooding and the Sundarbans*. Paris, France: Organisation for Economic Co-operation and Development.
18. Thomalla, F., Cannon, T., Huq, S., Klein, R.J.T., & Schaerer, C. (2010). Mainstreaming adaptation to climate change in coastal Bangladesh by

building civil society alliances. In L. Wallendorf, L. Ewing, S. Rogers, & C. Jones (eds), *Solutions to Coastal Disasters 2005* (pp. 668–684). Reston, VA: American Society of Civil Engineers.
19. Thomalla et al. (2010).
20. Reid, H., Simms, A., & Johnson, V. (2007). *Up in Smoke? Asia and the Pacific: The Threat from Climate Change to Human Development and the Environment.* London, UK: Working Group on Climate Change and Development.
21. Rawlani & Sovacool (2011).
22. Rawlani & Sovacool (2011).
23. Dasgupta, S., et al. (2010, November). Climate proofing infrastructure in Bangladesh: The incremental cost of limiting future inland monsoon flood damage. Policy Research Working Paper 5469. Washington, DC: World Bank.
24. Rawlani & Sovacool (2011).
25. Dasgupta et al. (2010) p. 4.
26. Ahmed, A.U. & Alam, M. (2010). Development of climate change scenarios with general circulation models. In S. Huq, Z. Karim, M. Asaduzzaman, & F. Mahtab (eds), *Vulnerability and Adaptation to Climate Change for Bangladesh* (pp. 13–20). London, UK: Kluwer Academic Publishers.
27. Agrawala et al. (2003).
28. Ahmed, A.U. (2008). *Assessment of Vulnerability to Climate Change and Adaptation Options for the Coastal People of Bangladesh.* Dhaka, Bangladesh: Practical Action Bangladesh.
29. World Bank (2000). Bangladesh: Climate change and sustainable development. Report No. 21104- BD, Rural Development Unit, South Asia Region. Dhaka, Bangladesh: World Bank.
30. Belt, D. (2011). Buoyant Bangladesh and the coming storm. *National Geographic*, 219(5), 58–83.
31. Nicholls, R.J., et al. (2007). Ranking port cities with high exposure and vulnerability to climate extremes. OECD Environment Working Paper No. 1 (ENV/WKP(2007)1). Paris, France: Organisation for Economic Co-operation and Development.
32. Agrawala et al. (2003).
33. Government of Bangladesh (2009). *Bangladesh Climate Change Strategy and Action Plan.* Dhaka, Bangladesh: MOEF.
34. Rawlani & Sovacool (2011).
35. UNFCCC (2010). National adaptation programs of action. Available at: http://unfccc.int/national_reports/napa/items/2719.php.
36. Government of Bangladesh (2005).
37. Government of Bangladesh (2005).
38. Feldman, S. & Geisler, C. (2012) Land expropriation and displacement in Bangladesh. *The Journal of Peasant Studies*, 39(3–4), 971–993.
39. Adnan, S. (2013). Land grabs and primitive accumulation in deltaic Bangladesh: Interactions between neoliberal globalization, state interventions, power relations and peasant resistance. *The Journal of Peasant Studies*, 40(1), 87–128.
40. Feldman, S. & Geisler, C. (2011). Land grabbing in Bangladesh: In-situ displacement of peasant holdings. Paper presented at the International Conference on Global Land Grabbing at the Institute of Development Studies, University of Sussex, (April).

41. Adnan, S. (2010). Resistance to accumulation by dispossession under neoliberal capitalism and globalization. Paper presented at the International Conference on Global Land Grabbing at the Institute of Development Studies, University of Sussex.
42. Feldman & Geisler (2012).
43. Shamsuddoha, M., Ullah, M.S., & Shahjahan, M. (2014). Land acquisition for climate-displaced communities in Bangladesh. In Scott Leckie (ed.), *Land Solutions for Climate Displacement* (pp. 159–194). London, UK: Routledge.
44. Barkat, A., Ara, R., Taheruddin, M., Hoque, S., & Islam, N. (2007). *Towards a Feasible Land Use Policy of Bangladesh*. Dhaka, Bangladesh: Human Development Research Centre (HDRC) and the Association for Land Reform and Development (ALRD), (August).
45. Barkat et al. (2007).
46. Quan, J. & Dyer, N. (2008). Climate change and land tenure: The implications of climate change for land tenure and land policy. Land Tenure Working Paper 2. Rome, Italy: Land and Water Division of the Food and Agriculture Organization (FAO) of the United Nations.
47. Islam, R. (2006). *Pre-and Post-tsunami Coastal Planning and Land-use Policies and Issues in Bangladesh*. Dhaka, Bangladesh: Integrated Coastal Zone Management Program.
48. Rashid, M.M. (2014). Political commitments and aspirations of grassroots coastal communities: A micro-level study in Bangladesh. *American Journal of Rural Development*, 2(2), 24–33.
49. Rashid, M.M. (2013). Lives and livelihoods of riverbank erosion displacees in Bangladesh: Need for protection framework. *Journal of Internal Displacement*, 3(1), 19–35.
50. Quan & Dyer (2008).
51. Raihan, S., Fatehin, S., & Haque, I. (2009). *Access to Land and Other Natural Resources by the Rural Poor: The Case of Bangladesh*. Dhaka, Bangladesh: South Asian Network on Economic Modeling (SANEM), Department of Economics, University of Dhaka.
52. Crow, B. & Murshid, K.A.S. (1994). Economic returns to social power: Merchants' finance and interlinkage in the grain markets of Bangladesh. *World Development*, 22(7), 1011–1030.
53. Lein, H. (2009). The poorest and most vulnerable? On hazards, livelihoods and labelling of riverine communities in Bangladesh. *Singapore Journal of Tropical Geography*, 30, 98–113.
54. Adnan (2010).
55. Huq, S. & Khan, M.R. Equity in national adaptation programs of action: The case of Bangladesh. In W.N. Adger, J. Paavola, S. Huq, & M.J. Mace (eds), *Fairness in Adaptation to Climate Change* (pp. 181–200). Cambridge, MA: MIT Press.
56. Quan & Dyer (2008).
57. Roy, R.C. (2000). Land rights of the indigenous peoples of the Chittagong Hill Tracts, Bangladesh. *IWGIA Document No. 99*. Copenhagen, Denmark: International Work Group for Indigenous Affairs.
58. Kabeer, N. (2011). Citizenship narratives in the face of bad governance: The voices of the working poor in Bangladesh. *The Journal of Peasant Studies*, 38(2), 325–353.

59. Blair, H. (2005). Civil society and pro-poor initiatives in rural Bangladesh: Finding a workable strategy. *World Development*, 33(6), 921–936.
60. Zaman, M.Q. (1991). Social structure and process in char land settlement in the Brahmaputra-Jamuna floodplain. *Man*, 26(4), 673–690.
61. Sarwar, M.G., Islam, R., & Monzoor, S. (2007). *Women's Rights to Land in Bangladesh: Roles, Limitations and Transformation*. Dhaka, Bangladesh: Unnayan Onneshan.
62. Islam, R. (2006). *Pre-and Post-tsunami Coastal Planning and Land-use Policies and Issues in Bangladesh, Integrated Coastal Zone Management Program*. Dhaka, Bangladesh: FAO.
63. Dasgupta et al. (2010).
64. World Bank (2013). *Resettlement action plan. Bangladesh*. Available at: http://documents.worldbank.org/curated/en/2013/02/17364117/bangladesh-first-phase-coastal-embankment-improvement-project-resettlement-plan.
65. Feldman & Geisler (2011) p. 7.
66. Islam (2006).
67. Quan & Dyer (2008).
68. Kartiki, K. (2011). Climate change and migration: A case study from rural Bangladesh. *Gender & Development*, 19(1), 23–38.
69. Kartiki (2011).
70. Islam (2006).
71. Adnan (2010).
72. Adnan, S. (2013) Land grabs and primitive accumulation in deltaic Bangladesh: Interactions between neoliberal globalization, state interventions, power relations and peasant resistance. *The Journal of Peasant Studies*, 40(1), 87–128.
73. Adnan (2013).
74. Paprocki, K. & Cons, J. (2014). Life in a shrimp zone: Aqua- and other cultures of Bangladesh's coastal landscape. *The Journal of Peasant Studies*, 41(6), 1109–1130.
75. Paprocki & Cons (2014).
76. Paprocki & Cons (2014).
77. Lein (2009).
78. Karim, A.H.M.Z. (2014). Flood and riverbank erosion displacees: Their indigenous survival strategies in two coastal villages in Bangladesh. *Asian Social Science*, 10(4, 22–33.
79. Feldman & Geisler (2012).
80. Mamun, M.Z. (1996). *Awareness, Preparedness, and Adjustment Measures of Riverbank Erosion Prone People: A Case Study for Settlement Planning for Disaster Mitigation in Bangladesh*. Dhaka, Bangladesh: Institute of Business Administration.
81. Parvin, G.A., Takahashi F., & Shaw, R. (2008). Coastal hazards and community-coping methods in Bangladesh. *Journal of Coastal Conservation*, 12(4), 181–193.
82. Parvin, Takahashi, & Shaw (2008).
83. Parvin, Takahashi, & Shaw (2008).
84. Haque, C.E. & Zaman, M.Q. (1989). Coping with riverbank erosion hazard and displacement in Bangladesh: Survival strategies and adjustments. *Disasters*, 13(4), 300–314.

85. Haque & Zaman (1989).
86. Feldman & Geisler (2012).
87. Haque & Zaman (1989).
88. Indra, D. (2000). Not just dis-placed and poor: How environmentally forced migrants in rural Bangladesh recreate space and place under trying conditions. In E.M. Gozdziak & D. Shandy (eds), *Rethinking Refuge and Displacement: Selected Papers of Refugees and Immigrants: Volume VIII* (pp. 163–191). Washington, DC: American Anthropological Association.
89. This mini-case study comes from personal correspondence with Mamun Rashid, February 8, 2015, and can be shared with readers who would like to see the two-page document.
90. Adnan (2011).
91. Kartiki (2011).
92. Sen, B. (2003). Drivers of escape and descent: Changing household fortunes in rural Bangladesh. *World Development*, 31(3), 513–534.

3 Degraded Seascapes: The Political Economy of the Dutch Delta Works

1. Hinkel, J. et al. (2014). Coastal flood damage and adaptation costs under 21st century sea-level rise. *PNAS*, 111(9), 3292–3297.
2. Fang, J., Sun, S., Shi, P., & Wang, J. (2014). Assessment and mapping of potential storm surge impacts on global population and economy. *International Journal of Disaster Risk Science*, 5(4), 323–331.
3. Fang et al. (2014).
4. Fang et al. (2014).
5. Kabat, P. et al. (2012). Climate proofing the Netherlands. *Nature*, 438, 283–284.
6. Tebaldi, C., Strauss, B.H., & Zervas, C.E. (2012). Modelling sea level rise impacts on storm surges along US coasts. *Environmental Research Letters*, 7(1), 014032.
7. Hoffman, R.N., et al. (2010). An estimate of increases in storm surge risk to property from sea level rise in the first half of the twenty-first century. *Weather, Climate, and Society*, 2, 271–293.
8. McRobie, A., Spencer, T., & Gerritsen, H. (2005). The big flood: North Sea storm surge. *Philosophical Transactions of the Royal Society A*, 363, 1263–1270.
9. McRobie et al. (2005).
10. Risk can be defined as "the probability of events and the magnitude of specific consequences. Risk is often defined as the multiplication of the two terms." See Kasperson et al. (1988). The social amplification of risk: A conceptual framework. *Risk Analysis*, 8(2), 177–187,177.
11. McRobie et al. (2005).
12. McRobie et al. (2005).
13. van Leussen, W. & Meijerink, S. (2014). *Environmentalism, Renewed Concern for Flood Safety, and the Europeanization of Dutch Water Policies, 1970–2010*. Alexandria, Virginia: Institute for Water Resources, US Army Corps of Engineers and Rijkswaterstaat, Ministry of Infrastructure and the Environment.
14. Kabat et al. (2005).

15. Jansen, O.E., Aarts, G.M., & Reijnders, P.J.H. (2013). Harbour porpoises Phocoena phocoena in the Eastern Scheldt: A resident stock or trapped by a storm surge barrier? *PLoS ONE*, 8(3), e56932.
16. Disco, C. (2002). Remaking "nature": The ecological turn in Dutch water management. *Science, Technology, & Human Values*, 27(2), 206–235.
17. Gerritsen, H. (2005). What happened in 1953? The big flood in the Netherlands in retrospect. *Philosophical Transactions of the Royal Society A*, 363(1831), 1271–1291.
18. Bijker, W.E. (2002). The Oosterschelde storm surge barrier. A test case for Dutch water technology, management, and politics. *Technology and Culture*, 43(3), 569–584.
19. Disco (2002).
20. Gerritsen (2005).
21. Gerritsen (2005).
22. Gerritsen (2005).
23. Gerritsen (2005).
24. Gerritsen (2005).
25. Gerritsen (2005).
26. Watson, I. & Finkl, C.W., Jr. (1990). State of the art in storm-surge protection: The Netherlands Delta project. *Journal of Coastal Research*, 6(3), 739–764.
27. See Disco, 2002, as well as d'Angremond, K. & Kooman, D. (1986). Eastern Scheldt storm surge barrier: Management aspects of a multibillion dollar project. *International Journal of Project Management*, 4(3), 149–157.
28. Bijker, W.E. (1993). Dutch, dikes and democracy. An argument against democratic, authoritarian, and neutral technologies. *Technology Assessment Texts No. 11*. Lyngby, Denmark: Unit of Technology Assessment, Technical University of Denmark.
29. d'Angremond & Kooman (1986); Disco (2002).
30. The official estimate is $7 billion in $1997, which we have updated here to $2014.
31. Smies, M. & Huiskes, A.H.L. (1981). Holland's Eastern Scheldt estuary barrier scheme: Some ecological considerations. *Ambio*, 10(4), 158–165.
32. Parma, S. (1978). Political aspects of the closure of the Eastern Scheldt estuary. *Hydrobiological Bulletin*, 12(3–4), 163–175.
33. Robin, J. (2013). Fighting the tide. *The New York Times*, February 24.
34. Delta Works Online Foundation/Deltawerken. (2004). *The Oosterschelde storm surge barrier*. Available at: http://www.deltawerken.com/The-Oosterschelde-storm-surge-barrier/324.html.
35. Leemans, A. (1986). Information as a factor of power and influence: Policymaking on the Delta/Oosterschelde Sea Defense Works. *Science Communication*, 8(1), 39–58.
36. Robin (2013).
37. Romijn, E. & Schutte, de Roon, J.C. (eds) (1995). *Netherlands Experiences With Integrated Water Management: Considerations for International Cooperation*. Delft, Netherlands: Rijkswaterstaat.
38. Bijker (1993).
39. Nienhuis, P.H. & Smaal, A.C. (1994). The Oosterschelde estuary, a case-study of a changing ecosystem: An introduction. *Hydrobiologia*, 282/283, 1–14
40. d'Angremond & Kooman (1986).

41. Leemans (1986).
42. Watson & Finkl Jr. (1990).
43. Delta Works Online Foundation/Deltawerken (2004).
44. Nelissen, H.A.M. (1986). Building on quicksand: The foundations of the Eastern Scheldt barrier. *Endeavour*, 10(3), 139–144.
45. d'Angremond & Kooman (1986).
46. Van der Tol, M.W.M. & Scholten, H. (1992). Response of the Eastern Scheldt ecosystem to a changing environment. *Netherlands Journal of Sea Research*, 30, 175–190.
47. Nelissen (1986).
48. Watson & Finkl Jr. (1990).
49. van Noortwijka, J.M. & Klatter, H.E. (1999). Optimal inspection decisions for the block mats of the Eastern-Scheldt Barrier. *Reliability Engineering and System Safety*, 65, 203–211.
50. These figures for overruns are quoted in Bijker and Aibar Punetes (1992), and could not be independently verified. Bijker, W. & Aibar Puentes, E. (1992). Dutch, dikes and democracy. An argument against democratic, flexible, good and bad technologies. *Proceedings of the 3rd European Congress on Technology Assessment*, November (pp. 538–557). Copenhagen: TeknologiNaevnet.
51. Robin (2013).
52. Disco (2002) p. 216.
53. Delta Works Online Foundation/Deltawerken (2004).
54. Watson & Finkl Jr. (1990).
55. Corvers, R. (2009). Vulnerable to flooding? Nature development and "room for the river": A governance perspective. In U.F. Paleo (ed.), *Building Safer Communities: Risk Governance, Spatial Planning, and Responses to Natural Hazards* (pp. 125–135). Amsterdam, the Netherlands: IOS Press.
56. Kabat et al. (2005).
57. d'Angremond & Kooman (1986).
58. US Army Corps of Engineers (2007). News release, February 2.
59. Barnett, J. & Hill, K. (2007). Design for rising sea levels. *Harvard Design Magazine*, p. 27. Available at: http://www.harvarddesignmagazine.org/issues/27/design-for-rising-sea-levels.
60. Brown, H. (2014). *Next Generation Infrastructure: Principles for Post-industrial Public Works*. Washington, DC: Island Press.
61. Bijker (2002).
62. Bijker (2002).
63. Disco (2002).
64. Bijker (1993).
65. Leemans (1986).
66. Bijker (1993).
67. Bijker (1993).
68. Knoester, M., Visser, J., Bannink, B.A., Colijn, C.J., & Broeders, W.P.A. (1983). The Eastern Scheldt Project. *Water Science & Techology*, 16(1–2), 51–77.
69. Bijker (1993).
70. Bijker (1993).
71. These were: Ballast Nedam, Boskalis Westminster, Baggermaatschappij Breejenhout, Hollandse Aanneming Maatschappij, Hollandse Beton Maatschappij, Van Oord-Utrecht, Stevin Baggeren, Stevin Beton en

Waterbouw, Adriaan Volker Baggermaatschappij, Adriaan Volker Beton en Waterbouw and Aannemerscombinatie Zinkwerken. See Delta Works Online Foundation/Deltawerken (2004).
72. Disco (2002).
73. Leemans (1986).
74. Disco (2002).
75. Disco (2002) p. 215.
76. Disco (2002) p. 215.
77. Meijerink, S. (2005). Understanding policy stability and change. The interplay of advocacy coalitions and epistemic communities, windows of opportunity, and Dutch coastal flooding policy 1945–2003. *Journal of European Public Policy*, 12(6), 1060–1077.
78. Meijerink (2005).
79. Leemans (1986).
80. Smies & Huiskes (1981).
81. Meijerink (2005).
82. Roth, D. & Warner, J. (2007). Flood risk, uncertainty and changing river protection policy in the Netherlands: The case of "calamity polders." *Tijdschrift voor Economische en Sociale Geografie*, 98(4), 519–525.
83. Parma (1978).
84. Disco (2002) p. 211.
85. Nienhuis & Smaal (1994).
86. van der Brugge, R., Rotmans, J., & Loorbach, D. (2005). The transition in Dutch water management. *Regional Environmental Change*, 5(4), 164–176.
87. Bijker (2002).
88. Leemans (1986).
89. Watson & Finkl Jr. (1990) p. 739.
90. Catlett, L., Wildhorn, S., Stanton, R.E., Roos, A., & Al, J. (1979). Controlling the Oosterschelde storm-surge barrier. Vol. I, Summary Report. RAND Corporation, Report R-2444/1-NETH. Santa Monica, CA: RAND.
91. Parma (1978).
92. Knoester et al. (1983).
93. Peet, G. & Gubbay, S. (1990). Marine protected areas in the North Sea. *International Journal of Estuarine & Coastal Law*, 5, 241–251.
94. This paragraph synthesizes the arguments made by Elgershuizen, J.H.B.W. (1981). Some environmental impacts of a storm surge barrier. *Marine Pollution Bulletin*, 12(8), 265–271; Eelkema, M., Wang, Z.B., & Stive, M.J.F. (2012). Impact of back-barrier dams on the development of the ebb-tidal Delta of the Eastern Scheldt. *Journal of Coastal Research*, 28(6): 1591–1605; Smaal, A.C. & Nienhuis, P.H. (1992). The eastern Scheldt (the Netherlands), from an estuary to a tidal bay: A review of responses at the ecosystem level. *Netherlands Journal of Sea Research*, 30, 161–173; Van der Tol, M.W.M. & Scholten, H. (1992). Response of the Eastern Scheldt ecosystem to a changing environment. *Netherlands Journal of Sea Research*, 30, 175–190; Leeuwdrent, W.J. (2012). Decision alternatives for the safety of the Eastern Scheldt: Will it be cost-effective to remove the Eastern Scheldt storm surge barrier in case of sea level rise? (unpublished thesis). Delft, the Netherlands: Delft University of Technology.
95. Elgershuizen (1981).
96. Parma (1978).

97. Wetsteyn, L.P.M.J. & Kromkamp, J.C. (1994). Turbidity, nutrients and phytoplankton primary production in the Oosterschelde (the Netherlands) before, during and after a large-scale coastal engineering project (1980–1990). *Hydrobiologia*, 282/283, 61–78.
98. Smies & Huiskes (1981).
99. Oenema, O. (1990b). Sulfate reduction in fine-grained sediments in the Eastern Scheldt, Southwest Netherlands. *Biogeochemistry*, 9(1), 53–74; Dame, R., Dankers, N., Prins, T., Jongsma, H., & Smaal, A. (1991). The influence of mussel beds on nutrients in the Western Wadden Sea and Eastern Scheldt Estuaries. *Estuaries*, 14(2), 130–138.
100. Oenema, O. (1990a). Pyrite accumulation in salt marshes in the Eastern Scheldt, Southwest Netherlands.*Biogeochemistry*, 9(1), 75–98.
101. Van den Broeck, H., De Wolf, H., Backeljau, T., & Blust, R. (2007). Effects of environmental stress on the condition of Littorina littorea along the Scheldt estuary (the Netherlands). *Science of the Total Environment*, 376, 346–358.
102. Jansen, Aarts & Reijnders (2013).
103. Smaal & Nienhuis (1992).
104. Hummel, H., Brummelhuis, E.B.M., & De Wolf, L. (1986). Effects on the benthic fauna of embanking an intertidal flat area. *Netherlands Journal of Sea Research*, 20(4), 397–406.
105. Eelkema, M. (2013). *Eastern Scheldt Inlet morphodynamics* (unpublished thesis, p. ix). Haveka, the Netherlands: Delft University of Technology.
106. Royal Society for the Protection of Birds. (2010). *Severn tidal power: RSPB briefing on the implications of the Eastern Scheldt (Oosterschelde) for the Severn Estuary*. Available at: http://www.rspb.org.uk/Images/RSPBbriefEasterScheldtreportfinal_tcm9-240984.pdf.
107. Bijker (1993).
108. Van den Broeck et al. (2010).
109. Meijerink (2005).
110. van der Brugge, Rotmans, & Loorbach (2005).
111. Commissie WB21 (Commissie Waterbeheer 21e eeuw). (2000). *Waterbeleid voor de 21e eeuw: Geef water de ruimte en de aandacht die het verdient [Water Policy for the 21st Century: Give Water the Space and Attention It Deserves]*. The Hague: Netherlands: Commissie WB21.
112. Kabat et al. (2005).
113. Roth & Warner (2007).
114. Roth, D. & Warner, J. (2009). Rural solutions for threats to urban areas: The contest over calamity polder. *Built Environment*, 35(4), 545–562.
115. Kabat et al. (2005).
116. Leeuwdrent (2012).
117. Liebowitz, S. J. & Margolis, S.E. (1995). Path dependence, lock-in, and history. *Journal of Law Economics & Organization*, 11(1), 205–226.
118. Unruh, G.C. (2000). Understanding carbon lock-in. *Energy Policy*, 12(1), 817–830; see also Spinardi, G. (2015). Up in the air: Barriers to greener air traffic control and infrastructure lock-in in a complex socio-technical system. *Energy Research & Social Science*, 6, 41–49.
119. Hughes, T.P. (1983). *Networks of Power: Electrification in Western Society, 1880–1930*. Baltimore, MD: Johns Hopkins University Press; Hughes, T.P. (1987). The evolution of large technological systems. In W.E. Bijker, T.P. Hughes, &

T. Pinch (eds), *The Social Construction of Technological Systems: New Directions in the Sociology and History of Technology*. Cambridge, MA: MIT Press.
120. McRobie et al. (2005).

4 Bloated Bodies: The Political Economy of Hurricane Katrina Recovery

1. Cutter, S.L., et al. (2006). The long road home: Race, class, and recovery from Hurricane Katrina. *Environment*, 48(2), 8–20.
2. Cutter et al. (2006).
3. Atkins, D. & Moy, E.M. (2005). Left behind: The legacy of Hurricane Katrina: Hurricane Katrina puts the health effects of poverty and race in plain view. *British Medical Journal*, 331(7522), 916–918; Lowe, J.S. & Shaw, T.C. (2009). After Katrina: Racial regimes and human development barriers in the Gulf Coast region. *American Quarterly*, 61(3), 80–827.
4. Bergal, J., et al. (eds). (2007). *City Adrift: New Orleans Before and After Katrina*. Baton Rouge, Louisiana: Louisiana State University Press/Center for Public Integrity.
5. Seed. R. (2007). Re: New Orleans, Hurricane Katrina, and the soul of the profession [Letter from Ray Seed, Professor of Civil and Environmental Engineering to William Marcuson III, President, American Society of Civil Engineers] October 30. Available at: http://aquadoc.typepad.com/files/katrina_and_soul_of_the_profession.pdf.
6. Bullard, R.D. & Wright, B. (2009a). Introduction. In R.D. Bullard & B. Wright (eds), *Race, Place, and Environmental Justice After Hurricane Katrina: Struggles to Reclaim, Rebuild, and Revitalize New Orleans and the Gulf Coast* (pp. 1–13, 1). Philadelphia, PA: Westview Press.
7. Elliott, J.R. & Pais, J. (2006). Race, class, and Hurricane Katrina: Social differences in human responses to disaster. *Social Science Research*, 35(2), 295–321.
8. Eysberg, C.D. (1989). The origins of the American urban system: Historical accident and initial advantage. *Journal of Urban History*, 15(2), 185–195.
9. Bullard, R.D. & Wright, B. (2009b). Race, place, and the environment in post-Katrina New Orleans. In R.D. Bullard & B. Wright (eds), *Race, Place, and Environmental Justice After Hurricane Katrina: Struggles to Reclaim, Rebuild, and Revitalize New Orleans and the Gulf Coast* (pp. 19–47). Philadelphia, PA: Westview Press.
10. Baker, C.R. (2014). Breakdowns of accountability in the face of natural disasters: The case of Hurricane Katrina. *Critical Perspectives on Accounting*, 25(7), 620–632.
11. Eysberg (1989).
12. Bullard & Wright (2009b).
13. Campanella R. (2007). An ethnic geography of New Orleans. *Journal of American History*, 94704–715; see also Glasmeier, A. & Leichenko, R. (2000). From Free Market Rhetoric to Free Market Reality: The Future of the US South in an Era of Globalization. In R. Tardanico & M.B. Rosenberg (eds), *Poverty or Development? Global Restructuring and Regional Transformations in the US South and the Mexican South* (pp. 19–39). New York: Routledge.
14. Campanella (2007).

15. Kates, R.W., Colten, C.E., Laska, S., & Leatherman, S.P. (2006). Reconstruction of New Orleans after Hurricane Katrina: A research perspective. *Proceedings of the National Academy of Sciences*, 103(40), 14653–14660.
16. Inniss, L.B. (2007). A domestic right of return? Race, rights, and residency in New Orleans in the aftermath of Hurricane Katrina. *Boston College Third World Law Journal*, 27(2), 325–373; Frey, W.H. & Singer, A. (2006). *Katrina and Rita Impacts on Gulf Coast Populations: First Census Findings*. Washington, DC: Brookings Institute.
17. Kates et al. (2006).
18. Henkel, K.E., Dovidio, J.F., & Gaertner, S.L. (2006). Institutional discrimination, individual racism, and Hurricane Katrina. *Analyses of Social Issues and Public Policy*, 6(1), 99–124.
19. Masozera, M., Bailey, M., & Kerchner, C. (2007). Distribution of impacts of natural disasters across income groups: A case study of New Orleans. *Ecological Economics*, 63, 299–306.
20. Wright, B. & Bullard, R.D. (2009a). Preface. In R.D. Bullard & B. Wright (eds), *Race, Place, and Environmental Justice After Hurricane Katrina: Struggles to Reclaim, Rebuild, and Revitalize New Orleans and the Gulf Coast* (pp. xix–xxii). Philadelphia, PA: Westview Press.
21. McDougall, H.A. (2007–2008). Hurricane Katrina: A story of race, poverty, and environmental injustice. *Howard Law Journal*, 51, 533–564.
22. McDougall (2007–2008).
23. Bullard & Wright (2009a, 2009b).
24. Masozera, Bailey & Kerchner (2007).
25. Peterson, C.H., et al. (1996). Ecological consequences of environmental perturbations associated with offshore hydrocarbon production: A perspective on the long-term exposures in the Gulf of Mexico. *Canadian Journal of Fisheries and Aquaculture Science*, 53, 2637–2654.
26. Silva, P. (2001). National Energy Policy. Hearing Before the House Subcommittee on Energy and Air Quality, February 18. Washington DC: US House of Representatives, pp. 113–116.
27. Bullard & Wright (2009a).
28. Baker (2014).
29. Driesen, D.M., et al. (2005). An unnatural disaster: The aftermath of Hurricane Katrina. *Center for Progressive Reform (CPR) Publication #512*, September. Washington DC: CPR.
30. McQuaid, J. & Schleifstein, M. (2002). Evolving danger. *Times Picayune*, June 23 (New Orleans), J12.
31. Weber, L. & Messias, D.K.H. (2012). Mississippi front-line recovery work after Hurricane Katrina: An analysis of the intersections of gender, race, and class in advocacy, power relations, and health. *Social Science & Medicine*, 74(11), 1833–1841.
32. Weber, L. (2014). Powering an unequal recovery. In S. Cutter, C.T. Emrich, J.T. Mitchell, W.W. Piegorsch, M.M. Smith, & L. Weber (eds), *The Recovery Divide: Hurricane Katrina and the Forgotten Coast of Mississippi* (pp. 90–113). New York, NY: Cambridge University Press.
33. Merdjanoff, A.A. (2013). There's no place like home: Examining the emotional consequences of Hurricane Katrina on the displaced residents of New Orleans. *Social Science Research*, 42(5), 1222–1235.

34. Driesen et al. (2005).
35. Kates et al. (2006).
36. Baker (2014).
37. Baker (2014).
38. Henkel, Dovidio, & Gaertner (2006).
39. Baker (2014).
40. Stivers, C. (2007). So poor and so black: Hurricane Katrina, public administration, and the issue of race. *Public Administration Review*, 67(s1), 48–56.
41. Baker (2014).
42. Stivers (2007).
43. Stivers (2007).
44. Baker (2014).
45. McDougall (2007–2008).
46. Henkel, Dovidio, & Gaertner (2006).
47. Stivers (2007).
48. Edwards, C. (2014). Why ending FEMA will improve disaster response. *Daily Caller*, November 18.
49. Weber & Messias (2012).
50. Federal Emergency Management Agency. (2005). Available from: https://www.fema.gov/news-release/2005/12/06/numbers-first-100-days.
51. Bullard & Wright (2009b).
52. Bullard & Wright (2009b).
53. Bullard & Wright (2009b).
54. Kates et al. (2006).
55. Elliott & Pais (2006) p. 302.
56. Cutter, S., Emrich, C.T., Mitchell, J.T., Piegorsch, W.W., Smith, M.M., & Weber, L. (eds). (2014). *The Recovery Divide: Hurricane Katrina and the Forgotten Coast of Mississippi*. New York, NY: Cambridge University Press.
57. Cutter et al. (2006).
58. Baker (2014).
59. See Tierney, K. (2012). Critical disjunctures: Disaster research, social inequality, gender, and Hurricane Katrina. In E. David & E. Enarson (eds), *The Women of Katrina: How Gender, Race, and Class Matter in an American Disaster* (pp. 245–258). Nashville, TN: Vanderbilt University Press; see Weber, L. (2013). Engendering disaster: Lessons from Hurricane Katrina. *Contemporary Sociology*, 42(5), 732–735.
60. Bullard & Wright (2009a, 2009b).
61. Bullard & Wright (2009a, 2009b).
62. Weber & Messias (2012).
63. Stivers (2007).
64. Stivers (2007).
65. Driesen et al. (2005).
66. King, R.J. (2009). Post-Katrina profiteering: The new big Easy. In R.D. Bullard & B. Wright (eds), *Race, Place, and Environmental Justice After Hurricane Katrina: Struggles to Reclaim, Rebuild, and Revitalize New Orleans and the Gulf Coast* (pp. 169–182). Philadelphia, PA: Westview Press.
67. King (2009) p. 169.
68. King (2009) p. 178.

69. Weber, L. & Peek, L. (2012). Documenting displacement: An introduction. In L. Weber & L. Peek (eds), *Displaced: Life in the Katrina Diaspora* (pp. 1–20). Austin, TX: University of Texas Press.
70. Inniss (2007).
71. Pardee, J.W. (2012). Living through displacement: Housing insecurity among low-income evacuees. In L. Weber & L. Peek (eds), *Displaced: Life in the Katrina Diaspora* (pp. 63–78, 64). Austin, TX: University of Texas Press.
72. Pardee (2012).
73. McFarlane, A. (2007–2008). Fighting for the high ground: Race, class, markets and the development done right in Post Katrina Recovery. *Washington & Lee Journal of Civil Rights & Social Justice*, 14, 77–96.
74. See Nguyen, T. (2007). A game of monopoly. *COLORLINES*, 17, May/June. Available at: http://www.colorlines.com/articles/game-monopoly
75. Nguyen (2007).
76. Lowe, J.S. (2012). Policy versus politics: Post-Hurricane Katrina lower-income housing restoration in Mississippi. *Housing Policy Debate*, 22(1), 57–73.
77. Lowe & Shaw (2009).
78. Wright, B. & Bullard, R.D. (2009b). Afterword: Looking back to move forward. In R.D. Bullard & B. Wright (eds), *Race, Place, and Environmental Justice After Hurricane Katrina: Struggles to Reclaim, Rebuild, and Revitalize New Orleans and the Gulf Coast* (pp. 265–274). Philadelphia, PA: Westview Press.
79. Long, A.P. (2007). Poverty is the new prostitution: Race, poverty, and public housing in Post-Katrina New Orleans. *The Journal of American History*, 94(3), 795–803.
80. Cutter et al. (2006).
81. Bullard, R.D., Johnson, G.S., & Torres, A.O. (2009). Transportation Matters: Stranded On the Side of the Road Before and After Disasters Strike. In R.D. Bullard & B. Wright (eds), *Race, Place, and Environmental Justice after Hurricane Katrina: Struggles to Reclaim, Rebuild, and Revitalize New Orleans and the Gulf Coast* (pp. 63–85). Philadelphia, PA: Westview Press.
82. Long (2007).
83. Henkel, Dovidio, & Gaertner (2006).
84. Bullard & Wright (2009a, 2009b).
85. Lacy, D.A. (2006–2007). The aftermath of Katrina: Race, undocumented workers, and the color of money. *Texas Wesleyan Law Review*, 13, 497–514.
86. Lacy (2006–2007) p. 506.
87. Lacy (2006–2007).
88. Lacy (2006–2007).
89. Quoted in Lacy (2006–2007).
90. Weber & Messias (2012).
91. McFarlane (2007–2008).
92. Inniss (2007).
93. McDougall (2007–2008).
94. Bullard & Wright (2009a, 2009b).
95. Masozera, Bailey & Kerchner (2007).
96. Bullard & Wright (2009a, 2009b).
97. Inniss (2007).

98. Olson, E.D. (2005). *The Environmental Effects of Hurricane Katrina*, October 6. Washington, DC: Natural Resources Defense Council.
99. Moe, T.L. (2010). Cleanup after Katrina: An analysis on policy, process, priorities, problems, and politics. *Disaster Prevention and Management: An International Journal*, 19(3), 314–332.
100. Mitsch, W.J. (2014). When will ecologists learn engineering and engineers learn ecology? *Ecological Engineering*, 65, 9–14.
101. DesRoches, R. (ed.). (2012). Hurricane Katrina: Performance of Transportation Systems. Reston, VA: American Society of Civil Engineers.
102. Poe, T. (2005). *Congressman Poe announces funding for Sabine-Neches Waterway* (press release), November 2. Available at: http://poe.house.gov/press-releases?ID=E0A7494A-B24A-44B9–93DC-0E351CA900A2.
103. Wright, B. (2007). *Testimony before the Subcommittee on Superfund and Environmental Health of the Senate Environment and Public Works Committee regarding Environmental Justice*, July 25. Washington, DC: US Government. Available at: http://www.epw.senate.gov/public/index.cfm?FuseAction=Files.View&FileStore_id=92ff035f-3a58–4baa-a5f8–4923f2597176.
104. Driesen et al. (2005).
105. Driesen et al. (2005).
106. Harden, M., Walker, N., & Akuno, K. (2007). Racial discrimination and ethnic cleansing in the United States in the aftermath of Hurricane Katrina: A report to the United Nations' Committee for the Elimination of Racial Discrimination, November 30. Geneva, Switzerland: Office of the United Nations High Commissioner for Human Rights (OHCHR).
107. US Environmental Protection Agency. (2006). *Summary results of sediment sampling in response to Hurricanes Katrina and Rita*, August 17. Washington DC: EPA. Available at: http://www.epa.gov/katrina/testresults/sediments/summary.html.
108. Wright (2007).
109. Natural Resources Defense Council. (2005). *Statement of Erik D. Olson Senior Attorney, Natural Resources Defense Council, Hurricane Katrina: Assessing the present environmental status*. Presented before the Subcommittee on Environment and Hazardous Materials of the Committee on Energy and Commerce. Washington DC: US House of Representatives.
110. Reed, A.L., Jr. (2005). The real divide. *The Progressive*, 69(11), 31.
111. Weber & Messias (2012).
112. Weber & Messias (2012).
113. Merdjanoff (2013).
114. Masozera, Bailey & Kerchner (2007).
115. Driesen et al. (2005).
116. Masozera, Bailey & Kerchner (2007).
117. Bullard & Wright (2009a, 2009b).
118. National Fair Housing Alliance (2005). No home for the holidays: Report on housing discrimination against Hurricane Katrina survivors, December. Washington, DC: National Fair Housing Alliance.
119. Fussell, E., Sastry, N., & VanLandingham, M. (2010). Race, socioeconomic status, and return migration to New Orleans after Hurricane Katrina. *Population and Environment*, 31(1–3), 20–42.

120. Bullard & Wright (2009b).
121. Elliott & Pais (2006).
122. Wright & Bullard (2009a) p. xx; Bullard & Wright (2009a, 2009b).
123. Stivers (2007).
124. Quoted in Henkel, Dovidio, & Gaertner (2006) p. 99.
125. Jones-DeWeever, A. (2007). Gender, race, and class: Disadvantage pre- and post-Katrina. *Women's Rights Law Reporter*, 28(1), 21–24.
126. Luft, R.E. (2008). Looking for common ground: Relief work in Post-Katrina New Orleans as an American parable of race and gender violence. *National Women's Studies Association Journal*, 20(3), 5–31.
127. David, E. & Enarson, E. (eds). (2012). *The Women of Katrina: How Gender, Race, and Class Matter in an American Disaster* (p. 7). Nashville, TN: Vanderbilt University Press.
128. Peek, L. & Fothergill, A. (2008). Displacement, gender, and the challenges of parenting after Hurricane Katrina. *National Women's Studies Association Journal*, 20(3), 69–105.
129. Tyler, P. (2007). The Post-Katrina, semiseparate world of gender politics. *The Journal of American History*, 94(3), 780–788.
130. David & Enarson (2012); see also Enarson, E. (2012). *Women Confronting Natural Disaster: From Vulnerability to Resilience*. Boulder, CO: Lynn Reiner Publishers.
131. Bullard & Wright (2009b).
132. Bullard & Wright (2009b) p. 38.
133. Peek, L. (2012). They call it "Katrina Fatigue": Displaced families and discrimination in Colorado. In L. Weber & L. Peek (eds), *Displaced: Life in the Katrina Diaspora* (pp. 31–46, 34). Austin, TX: University of Texas Press.
134. Fothergill, A. & Peek, L. (2012). Permanent temporariness: Displaced children in Louisiana. In L. Weber & L. Peek (eds), *Displaced: Life in the Katrina Diaspora* (pp. 119–143). Austin, TX: University of Texas Press.
135. Peek (2012).
136. Logan, J.R. (2009). Unnatural disaster: Social impacts and policy changes after Katrina. In R.D. Bullard & B. Wright (eds), *Race, Place, and Environmental Justice after Hurricane Katrina: Struggles to Reclaim, Rebuild, and Revitalize New Orleans and the Gulf Coast* (pp. 249–264). Philadelphia, PA: Westview Press.
137. Freudenburg, W.R., Gramling, R.B., Laska, S., & Erikson, K. (2009). *Catastrophe in the Making: The Engineering of Katrina and the Disasters of Tomorrow*. Washington, DC: Island Press.
138. Inniss (2007).
139. Inniss (2007).
140. Atkins & Moy (2005).
141. McFarlane (2007–2008).
142. Lowe & Shaw (2009).
143. Lowe & Shaw (2009).
144. Weber, L. (2012). When demand exceeds supply: Disaster response and the southern political economy. In L. Weber & L. Peek (eds), *Displaced: Life in the Katrina Diaspora* (pp. 79–103, 79). Austin, TX: University of Texas Press.
145. Masozera, Bailey, & Kerchner (2007).
146. Lein, L., Angel, R., Beausoliel, J., & Bell, H. (2012). The basement of extreme poverty: Katrina survivors and poverty programs. In L. Weber & L. Peek (eds),

Displaced: Life in the Katrina Diaspora (pp. 47–62). Austin, TX: University of Texas Press.

147. Weber (2012).
148. Kates et al. (2006).
149. Bullard & Wright (2009a, 2009b).
150. Laska, S. & Morrow, B.H. (2006). Social vulnerabilities and Hurricane Katrina: An unnatural disaster in New Orleans. *Marine Technology Society Journal*, 40(4), 16–26.
151. Bullard & Wright (2009a) p. 2; Bullard & Wright (2009b).
152. Freudenburg et al. (2009).
153. Mileti, D. (1999). *Disasters by Design: A Reassessment of Natural Hazards in the United States*. Washington, DC: National Academies Press.

5 The Perils of Climate Diplomacy: The Political Economy of the UNFCCC

1. *Recent Monthly Average Mauna Loa CO2*. Available at: http://www.esrl.noaa.gov/gmd/ccgg/trends/co2_data_mlo.html.
2. Antholis, W. & Talbott, S. (2010). *Fast Forward: Ethics and Politics in the Age of Global Warming*. Washington, DC: Brookings Institution Press.
3. Antholis & Talbott (2010).
4. Tomasello, M. (1999). *The Cultural Origins of Human Cognition*. Cambridge, MA: Harvard University Press; Tomasello, M. and Vaish, A. (2013). Origins of human cooperation and morality. *Annual Review of Psychology*, 64, 231–255.
5. Victor, D. (2011). *Global Warming Gridlock: Creating More Effective Strategies for Protecting the Planet*. Cambridge: Cambridge University Press.
6. UNFCCC (2006a). *Climate Change: Impacts, Vulnerabilities and Adaptation in Developing Countries*. Bonn, Germany: UNFCCC.
7. United Nations (1992a). *The Framework Convention on Climate Change*. A/AC.237/18 (Part II)/Add.1. New York, NY: United Nations.
8. United Nations (1992a) Article 4.1(f).
9. Selin, H. & Linnér, B.-O. (2005). The quest for global sustainability: International efforts on linking environment and development. CID Graduate Student and Postdoctoral Fellow Working Paper No. 5. Cambridge, MA: Science, Environment and Development Group, Center for International Development, Harvard University.
10. Frank, A. G., Chew, S. C., & Denemark, R. A. (1996). *The underdevelopment of development : essays in honor of Andre Gunder Frank*. Thousand Oaks: Sage Publications.
11. Linnér, B.O. & Selin, H. (2013). The United Nations Conference on Sustainable Development: Forty years in the making. *Environment and Planning C: Government and Policy*, 31(6), 971–987.
12. WMO (1979). *Proceedings of the World Climate Conference: A Conference of Experts on Climate and Mankind*. (p. 710).WMO, 537, February 12–23. Geneva: Switzerland: WMO.
13. Holdgate, M.W., Kassas, M., & White, G.F. (1982). *The World Environment 1972–1982*. Dublin, Ireland: Tycooly International Publishers/United Nations Environment Programme.

14. World Commission on Environment and Development (1987). *Our Common Future*, (para. 27). New York, NY: Oxford University Press.
15. United Nations (1989a). *Protection of Global Climate for Present and Future Generations of Mankind*. New York, NY: UN General Assembly.
16. South Pacific Forum (1989). *Final communique of the Twentieth South Pacific Forum*, July 10–11. Tarawa, Kiribati; Commonwealth Heads of Government Meeting (1989). *Langkawi Declaration on the Environment*, October 21. Langkawi, Malaysia; Conference of Non-Aligned Heads of State (1989). *Ninth Conference of Heads of State of the Non-Aligned States*. Belgrade, Serbia.
17. Linnér, B.-O. & Jacob, M. (2005). From Stockholm to Kyoto: A review of the globalization of global warming policy and north–south relations. *Globalizations*, 2(3), 403–415.
18. United Nations (1992b). INC5, Part I: Report, New York, 1992. A/AC.237/18 (Part I).
19. G-77. Caracas Declaration of the Ministers of Foreign Affairs of the Group of 77 on the Occasion of the Twenty-Fifth Anniversary of the Group. Caracas, Venezuela, Group of 77.
20. South Pacific Forum (1989).
21. Pielke, R., Prins, G., Rayner, S., & Sarewitz, D. (2007). Lifting the taboo on adaptation. *Nature, 445*(7128), 597–598; Schipper, L. (2006). Conceptual history of adaptation in the UNFCCC process. *RECIEL, 15*(1), 82–92.
22. Barrett, S. (2008). Climate treaties and the imperative of enforcement. *Oxford Review of Economic Policy, 24*(2), 239–258. DOI: 10.1093/oxrep/grn015; Schipper, L. (2006). Conceptual history of adaptation in the UNFCCC process. *RECIEL, 15*(1), 82–92.
23. United Nations (1989). *Possible Adverse Effects of Sea-level Rise on Islands and Coastal Areas, Particularly Low-lying Coastal Areas*. New York, NY: UN Secretary-General.
24. United Nations (1989). *United Nations Conference on Environment and Development*.
25. United Nations (1989a).
26. Miller, M.A.L. (1995). *The Third World in Global Environmental Politics*. Buckingham, UK: Open University Press.
27. United Nations (1992a) Article 4.8.
28. United Nations (1992a) Article 4.9.
29. UNFCCC (2012). COP18: Agreed outcome pursuant to the Bali Action Plan. Draft decision-/CP.18. Bonn, Germany: UNFCCC.
30. Friman, M. & Linnér, B.-O. (2008). Technology obscuring equity: Historical responsibilities in UNFCCC negotiations. *Climate Policy*, 8(4), 339–354.
31. Jagers, S.C. & Duus-Otterström, G. (2008). Dual climate change responsibility: On moral divergences between mitigation and adaptation. *Environmental Politics*, 17(4), 576–591; Dellink, R., et al. (2009). Sharing the burden of financing adaptation to climate change. *Global Environmental Change*, 19(4), 411–421; Friman, M. & Strandberg, G. (2014). Historical responsibility for climate change: Science and the science–policy interface. *WIREs Climate Change*, 5(3), 297–316, 325.
32. Friman & Strandberg (2014).
33. UNFCCC (2011) 1/CP.16.
34. Khan, M.R. & Roberts, J.T. (2013). Adaptation and international climate policy. *WIREs Climate Change*, 4(3), 171–189.

35. Huq, S., Roberts, E., & Fenton, A. (2013). Loss and damage. *Nature Climate Change*, 3(11), 947–949; Friman, M. & Linnér, B.-O. (2015). Perspectives on the Green Climate Fund: Possible Compromises on Capitalization and Balanced Allocation. *Climate and Development*; Agrawala, S. & van Aalst, M. (2008). Adapting development cooperation to adapt to climate change. *Climate Policy*, 8(2), 183–193.
36. Ghani, A., Lockhart, C., & Carnahan, M. (2005). Closing the sovereignty gap: An approach to state-building. Working Paper, No. 253. London, UK: Overseas Development Institute (ODI).
37. Wendt, A. (1999). *Social Theory of International Politics*. Cambridge, MA: Cambridge University Press; Young, O. (2002). *The Institutional Dimensions of Environmental Change: Fit, Interplay, and Scale*. Cambridge, MA: MIT Press.
38. United Nations Development Programme (2009). *Evaluation of UNDP Work with Least Developed Countries Fund and Special Climate Change Fund Resources*. New York, NY: UNDP Evaluation Office.
39. United Nations Development Programme (2009).
40. UNDP (2009).
41. UNDP (2009).
42. UNDP (2009).
43. Danish Ministry of Foreign Affairs (2010). Review of the follow up on the LDCF Evaluation and information update on the LDCF and SCCF. GEF/LDCF. SCCF.9/Inf.7, October 20. Copenhagen, Denmark: DMFA.
44. GEF Secretariat (2010). *Climate Change Forum, The Least Developed Countries Fund & the Special Climate Change Fund: Factsheet, 2010*. Washington, DC: GEF Secretariat.
45. Barnett, M. & Finnmore, M. (2004). *Rules for the World: International Organizations in Global Politics*. New York, NY: Cornell University Press; Linnér, B.-O. (2006). Authority through synergism: The roles of climate change linkages. *European Environment*, 16(5), 278–289.
46. See Paavola, J., Adger, W.N., & Huq, S. (2006). Multifaceted justice in adaptation to climate change. In W.N. Adger, J. Paavola, S. Huq, & M.J. Mace (eds), *Fairness in Adaptation to Climate Change* (pp. 263–277). Cambridge, MA: MIT Press; Roberts, J.T. & Parks, B.C. (2007). *A Climate of Injustice: Global Inequality, North–South Politics, and Climate Policy*. Cambridge, MA: MIT Press.
47. Schroeder, H., Boykoff, M.T., & Spiers, L. (2012). Equity and state representations in climate negotiations. *Nature Climate Change*, 2(12), 834–836; Roger, C. & Belliethathan, S. (2014). Africa in the global climate change negotiations. *International Environmental Agreements: Politics, Law and Economics*. DOI:10.1007/s10784–014–9244–7.
48. Gordon, R. (2007). Climate change and the poorest nations: Further reflections on global inequality. *University of Colorado Law Review*, 78, 1559–1624.
49. Gilligan, A. (2009). Copenhagen climate summit: 1,200 limos, 140 private planes and caviar wedges. *The Daily Telegraph*, December 5, London.
50. Schneider, S.H. & Lane, J. (2006). Dangers and thresholds in climate change and the implications for justice. In W.N. Adger, J. Paavola, S. Huq, & M.J. Mace (eds), *Fairness in Adaptation to Climate Change* (pp. 23–51). Cambridge, MA: MIT Press.

51. Shue, H. (1992). The unavoidability of justice. In A. Hurrell & B. Kingsbury (eds), *The International Politics of the Environment: Actors, Interests, and Institutions* (pp. 373–397). Oxford, UK: Oxford University Press.
52. Khan, M.R. (2014). *Toward a Binding Climate Change Adaptation Regime: A Proposed Framework* (p. 29). London, UK: Routledge.
53. Demeritt, D. (2001). The construction of global warming and the politics of science. *Annals of the Association of American Geographers*, 91(2), 307–337; Linnér & Jacob (2005); Khan (2014).
54. Gordon (2007).
55. Hjerpe, M. & Linnér, B.-O. (2010). Functions of COP side-events in climate change governance. *Climate Policy*, 10(2), 167–180; Nasiritousi, N., Hjerpe, M., & Linnér, B.-O. (2014). The role of non-state actors in climate change governance: Understanding agency through governance profiles. *International Environmental Agreements*. DOI: 10.1007/s10784-014-9243-8.
56. Martello, M.L. & Jasanoff, S. (2004). Globalization and environmental governance. In S. Jasanoff & M.L. Martello (eds), *Earthly Politics: Local and Global in Environmental Governance* (pp. 1–29). Cambridge, MA: MIT Press.
57. Phelps, J., Webb, E.L., & Agarwal, A. (2010). Does REDD+ threaten to recentralize forest governance? *Science*, 328 (5976), 312–313.
58. Frechette, A., de Bresser, M., & Hofstede, R. (2014). External evaluation of the United Nations collaborative programme on Reducing Emissions from Deforestation and Forest Degradation in developing countries (the UN–REDD+ programme, draft report). Available at: http://www.un-redd.org/Portals/15/documents/UN-REDD%20Evaluation%20Final%20Report%20Volume1%20June2014%20EN.pdf.
59. Phelps, Webb, & Agarwal (2010).
60. Frechette, de Bresser, & Hofstede (2014).
61. Phelps, Webb, & Agarwal (2010).
62. Forest Peoples Programme (2013). Safeguards in REDD+ financing schemes, April 29. Available at: http://www.forestpeoples.org/topics/forest-carbon-partnership-facility-fcpf/news/2013/04/safeguards-redd-financing-schemes.
63. Forest Peoples Programme (2013).
64. See Bernow, S., Kartha, S., Lazarus, M., & Page. T. (2001). Cleaner generation, free-riders, and environmental integrity: Clean development mechanism and the power sector. *Climate Policy*, 1(2), 229–249; as well as Lohmann, L. (2009). When markets are poison: Learning about climate policy from the financial crisis. *The Corner House Briefing 40*, September. Dorset, UK: Corner House.
65. Nayar, A. (2009). When the ice melts. *Nature*, 461 (7267), 1042–1046.
66. Sovacool, B.K. (2011). Conceptualizing hard and soft paths for climate change adaptation. *Climate Policy*, 11(4), 1177–1183.
67. Atela, J.O., Quinn, C.H., & Minang, P.A. (2014). Are REDD projects pro-poor in their spatial targeting? Evidence from Kenya. *Applied Geography*, 52, 14–24; UNFCCC (2011). Report of the Conference of the Parties on its seventeenth session, held in Durban from November 28 to December 11, 2011. Addendum. Durban, South Africa: UNFCCC; Campbell, B.M. (2009). Beyond Copenhagen: REDD+, agriculture, adaptation strategies and poverty. *Global Environmental Change*, 19(4), 397–399.

68. Najam, A., Huq, S., & Sokonae, Y. (2003). Climate negotiations beyond Kyoto: Developing countries' concerns and interests. *Climate Policy*, 3(3), 221–231.
69. Anand, R. (2004). *International Environmental Justice: A North–South Dimension*. Aldershot, UK: Ashgate; Gupta, J. (1997). The climate change issue. In J. Gupta (ed.), *The Climate Change Convention and Developing Countries: From Conflict to Consensus?* London, UK: Kluwer Academic.
70. Fankhauser, S. & Martin, N. (2010). The economics of the CDM levy: Revenue potential, tax incidence and distortionary effects. *Energy Policy*, 38(1), 357–363.
71. Chukwumerije, O. (2010). Climate justice and the international regime. *Wiley Interdisciplinary Reviews: Climate Change*, 1(3), 462–474.
72. Sterk, W., Luhmann, H.-J., & Mersmann, F. (2011). *How Much is 100 Billion US Dollars? Climate Finance Between Adequacy and Creative Accounting*. Berlin, Germany: Friedrich-Ebert-Stiftung.
73. Vidal, J. (2014). Rich countries "backsliding" on climate finance. *The Guardian*, November 18.
74. Buchner, B., Falconer, A., Hervé-Mignucci, M., & Trabacchi, C. (2012). *The Landscape of Climate Finance 2012*. San Francisco, CA: Climate Policy Initiative.
75. Buchner et al. (2012).
76. GEF (2009). *Accessing Resources Under the Least Developed Countries Fund*. Washington, DC: GEF.
77. Remling, E. & Persson, A. (2014). Who is adaptation for? Vulnerability and adaptation benefits in proposals approved by the UNFCCC Adaptation Fund. *Climate and Development*, 7(1), 16–34.
78. Remling & Persson (2014).
79. UNFCCC (1992) Article 4.1.
80. IPCC (2000); Markandya, A. & Galarraga, I. (2011). Technologies for adaptation: An economic perspective. In A. Olhoff & S.T.L. Christiansen (eds), *Technologies for Adaptation: Perspectives and Practical Experiences*. Roskilde, Denmark: UNEP Risø Centre on Energy, Climate and Sustainable Development; Biagini, B., Kuhl, L., Sims Gallagher, K. & Ortiz, C. (2014). Technology transfer for adaptation. *Nature Climate Change*, 4(9), 828–834.
81. Huq, S. & Rabbani, G. (2011). Adaptation technologies in agriculture: The economics of rice-farming technology in climate-vulnerable areas of Bangladesh. In A. Olhoff & S.T.L. Christiansen (eds), *Technologies for Adaptation: Perspectives and Practical Experiences*. Roskilde, Denmark: UNEP Risø Centre on Energy, Climate and Sustainable Development; De Lopez, T., Ponlok, T., & Dany, V. (2011). Community adaptation to climate change in Cambodia: Technology and development aspects for agriculture. In A. Olhoff & S.T.L. Christiansen (eds), *Technologies for Adaptation: Perspectives and Practical Experiences*. Roskilde, Denmark: UNEP Risø Centre on Energy, Climate and Sustainable Development.
82. UNFCCC (2007); Sharma, S. & Moehner, A. (2011), The evolution of "technologies for adaptation" in the international climate change negotiations. In A. Olhoff & S.T.L. Christiansen (eds), *Technologies for Adaptation: Perspectives and Practical Experiences*. Roskilde, Denmark: UNEP Risø Centre on Energy, Climate and Sustainable Development.
83. Hoekman, B.M., Maskus, K.E., & Saggi, K. (2005). Transfer of technology to developing countries: Unilateral and multilateral policy options. *World Development*, 33(10), 1587–1602.

84. Baumgartner, T., Burns, T.R., & Deville, P. (2015). *The Shaping of Socio-economic Systems: The Application of the Theory of Actor-system Dynamics to Conflict, Social Power, and Institutional Innovation in Economic Life*. New York, NY: Routledge.
85. Callaway, J.M. (2014). The role of technology in adaptation. In I. Galarraga, A. Markandya, & E. Sainz de Murieta (eds), *Routledge Handbook of the Economics of Climate Change Adaptation*. New York, NY: Routledge.
86. Jorgenson, A.K. & Givens, J.E. (2014). The emergence of new World-Systems perspectives on global environmental change. In D.A. Sonnefeld, S. Lockie, & D.R. Fisher (eds), *Routledge International Handbook of Social and Environmental Change*. New York, NY: Routledge.
87. Sovacool, B.K. (2008). Placing a glove on the invisible hand: How intellectual property rights may impede innovation in energy Research and Development (R&D). *Albany Law Journal of Science & Technology*, 18(2), 381–440.
88. International Energy Agency (IEA), International Renewable Energy Agency (IRENA), Food and Agricultural Organization (FAO), Renewable Energy Policy Network for the 21st Century (REN21), United Nations Environment Program (UNEP), World Bank (WB)/Energy Sector Management Assistance Program (ESMAP), Renewable Energy Chapter (2015). *Global Tracking Framework, 2015 Sustainable Energy for All* (SE4ALL), January 20. Washington, DC: IEA.
89. Sustainable and Renewable Energy Distribution Authority (SREDA) (2015). *RE development initiatives in Bangladesh*. Available at: http://sreda.gov.bd/index.php/re-development-initiatives-in-bangladesh.

6 Principles and Best Practices for Climate Change Adaptation

1. Ostrom, E. (2000). Collective action and the evolution of social norms. *Journal of Economic Perspectives*, 14(3), 137–158.
2. McCarthy, J. (2005). Commons as counterhegemonic projects. *Capitalism, Nature, Socialism*, 16(1), 9–24.
3. Dietz, T., Ostrom, E., & Stern, P. (2003). The struggle to govern the commons. *Science*, 302(5652), 1907–1912; Ostrom (2000).
4. These arguments do draw from (and extend) some of our earlier work, notably Brown, M.A. & Sovacool, B.K. (2011). *Climate Change and Global Energy Security: Technology and Policy Options*. Cambridge, MA: MIT Press; and Sovacool, B.K. (2011). An international comparison of four polycentric approaches to climate and energy governance. *Energy Policy*, 39(6), 3832–3844.
5. These arguments have also been adapted from some of our earlier research on energy justice, especially Sovacool, B.K. & Dworkin, M.H. (2014). *Global Energy Justice: Problems, Principles, and Practices*. Cambridge, UK: Cambridge University Press; Sovacool, B.K., Sidortsov, R., & Jones, B. (2014). *Energy Security, Equality, and Justice*. London, UK: Routledge; and Sovacool, B.K. (2013). *Energy & Ethics: Justice and the Global Energy Challenge*. New York, NY: Palgrave Macmillan.
6. Shue, H. (1992). The unavoidability of justice. In A. Hurrell & B. Kingsbury (eds), *The International Politics of the Environment: Actors, Interests, and Institutions* (pp. 373–397). Oxford, UK: Oxford University Press.

7. Beckerman, W. (1999). Sustainable development and our obligations to future generations. In A. Dobson (ed.), *Fairness and Futurity: Essays on Environmental Sustainability and Social Justice* (pp. 71–92). Oxford, UK: Oxford University Press.
8. Page, E.A. (2007). Intergenerational justice of what: Welfare, resources or capabilities? *Environmental Politics*, 16(3), 453–469.
9. Drawn from Hillman, M. (2004). The importance of environmental justice in stream rehabilitation. *Ethics, Place, and Environment*, 7(1/2), 2–19; and Konow, J. (2003). Which is the fairest one of all? A positive analysis of justice theories. *Journal of Economic Literature*, 41(4), 1188–1239.
10. It is based on two sub-principles: countries that have (1) historically significantly contributed to the present atmospheric concentrations of greenhouse gases and associated damages and (2) high financial and technological capacities are given greater responsibility to contribute to the goals of the Convention. United Nations (1992). *The Framework Convention on Climate Change*. A/AC.237/18 (Part II)/Add.1. New York, NY: United Nations.
11. Weston, B.H. (2008). Climate change and intergenerational justice: Foundational reflections. *Vermont Journal of Environmental Law*, 9, 375–430.
12. Rawls, R. (1970). Justice as fairness. In W. Sellars & J. Hospers (eds), *Readings in Ethical Theory* (pp. 578–595). Englewood Cliffs, NJ: Prentice-Hall.
13. Nussbaum, M.C. (2006). *Frontiers of Justice: Disability, Nationality, Species Membership*. Cambridge, MA: Belknap Press, pp. 9–10.
14. Rawls, J. (1999). *A Theory of Justice: Revised Edition*. Cambridge, MA: Belknap Press, p. 12.
15. Sandel, M. (2009). *Justice. What's The Right Thing To Do?* New York: Farrar, Straus & Giroux, p. 142.
16. Brighouse, H. (2004). *Justice*. Cambridge, UK: Polity Press, pp. 52–53.
17. Walker, G. (2012). *Environmental Justice: Concepts, Evidence, and Politics*. London, UK: Routledge.
18. Maslow, A. (1954/1987). Motivation and personality (3rd rev. edn). New York, NY: Harper & Row.
19. Wolpert, S. (2008). Brain reacts to fairness as it does to money and chocolate, study shows. *UCLA Newsroom*, April 21.
20. Brosnan, S.F. & de Waal, F.B.M. (2003). Monkeys reject unequal pay. *Nature*, 424(September 18), 297–299.
21. Adger, W.N., Paavola, J., & Huq, S. (2006). Toward justice in adaptation to climate change. In W.N. Adger, J. Paavola, S. Huq, & M.J. Mace (eds), *Fairness in Adaptation to Climate Change* (pp. 1–19). Cambridge, MA: MIT Press; Barry, B. (1995). *Justice as Impartiality*. Oxford: Oxford University Press; Salazar, D.J. & Alper, D.K. (2011). Justice and environmentalisms in the British Columbia and US Pacific Northwest environmental movements. *Society & Natural Resources*, 24(8), 767–784.
22. Weston (2008); see also Weston, B.H. & Bach, T. (2008). *Climate Change and Intergenerational Justice: Present Law, Future Law*. Royalton, VT: Vermont Law School.
23. Paavola, J., Adger, W.N., & Huq, S. (2006). Multifaceted justice in adaptation to climate change. In W.N. Adger, J. Paavola, S. Huq, & M.J. Mace (eds), *Fairness in Adaptation to Climate Change* (pp. 263–277). Cambridge, MA: MIT Press.

24. Dolan, P., Edlin, R., Tsuchiya, A., & Wailoo, A. (2007). It ain't what you do, it's the way that you do it: Characteristics of procedural justice and their importance in social decision-making. *Journal of Economic Behavior & Organization*, 64, 157–170.
25. Haggett, C. (2009). Public engagement in planning for renewable energy. In S. Davoudi, J. Crawford, & A. Mehmood (eds), *Planning for Climate Change: Strategies for Mitigation and Adaptation for Spatial Planners* (pp. 297–307). London, UK: Earthscan.
26. Gutmann, A. (2004). *Why Deliberative Democracy?* Princeton, NJ: Princeton University Press; Brown, M.B. (2009). *Science in Democracy: Expertise, Institutions, and Representation*. Cambridge, MA: MIT Press; Lövbrand, E., Pielke, R., et al. (2011). A democracy paradox in studies of science and technology. *Science, Technology & Human Values*, 36(4), 474–496.
27. Parkinson, J. (2003). Legitimacy problems in deliberative democracy. *Political Studies*, 51(1), 180–196, 181.
28. Dryzek, J. (2001). Legitimacy and economy in deliberative democracy. *Political Theory*, 29(5), 651–669; Eckersley, R. (2004). *The Green State: Rethinking Democracy and Sovereignty*. Cambridge, MA: MIT Press.
29. André, K., Simonsson, L. et al. (2012). Method development for identifying and analysing stakeholders in climate change adaptation processes. *Journal of Environmental Policy & Planning*, 14(3), 243–261.
30. Cooke, B. & Kothari, U. (2001). *Participation: The New Tyranny?* New York, NY: Zed Books.
31. Barbour, I.G. (1980). *Technology, Environment, and Human Values*. Westport, CT: Praegar, p. 204.
32. Lewis, S. (1999). The precautionary principle and corporate disclosure. In C. Raffensperger & J.A. Tickner (eds), *Protecting Public Health and the Environment: Implementing the Precautionary Principle* (pp. 241–251). Washington, DC: Island Press.
33. Ostrom, E. (2010). Polycentric systems for coping with collective action and global environmental change. *Global Environmental Change*, 20, 550–557; Sovacool, B.K. (2011). Conceptualizing hard and soft paths for climate change adaptation. *Climate Policy*, 11(4), 1177–1183.
34. Ostrom, E. (2009a). The governance challenge: Matching institutions to the structure of socio-ecological systems. In S. Levin (ed.), *The Princeton Guide to Ecology*. Trenton, NJ: Princeton University Press; Ostrom, E. (2009b). A polycentric approach for coping with climate change. Report Prepared for the WDR2010 Core Team, Development and Economics Research Group, World Bank. Bloomington, IN: Indiana University.
35. Ostrom, V., Tiebout, C., & Warren, R. (1961). The organization of government in metropolitan areas. *American Political Science Review*, 55, 831–842; Ostrom, V. & Ostrom, E. (1977). Public goods and public choices. In E.S. Savas (ed.), *Alternatives for Delivering Public Services: Toward Improved Performance* (pp. 7–49). Boulder, CO: Westview Press.
36. Andersson, K.P. & Ostrom, E. (2008). Analyzing decentralized resource regimes from a polycentric perspective. *Policy Sciences*, 41(1), 71–93.
37. Hooghe, L. & Marks, G. (2003). Unraveling the central state, but how? Types of multi-level governance. *American Political Science Review*, 97(2), 233–243.

38. Blomquist, W. (2009). Multi-level governance and natural resource management: The challenges of complexity, diversity, and uncertainty. In V. Beckmann & M. Padmanabhan (eds), *Institutions and Sustainability* (pp. 109–126). New York, NY: Springer.
39. Andersson & Ostrom (2008).
40. McGinnis, M. (2005). Beyond individuals and spontaneity. *Journal of Economic Behavior and Organization*, 57, 167–172.
41. Brock, G. (2009). *Global Justice: A Cosmopolitan Account*. Oxford: Oxford University Press.
42. International Council on Human Rights Policy. (2008). *Climate Change and Human Rights: A Rough Guide*. Versoix, Switzerland: ICHRP.
43. International Council on Human Rights Policy (2008).
44. Kolstad, I. & Wiig, A. (2009). Is transparency the key to reducing corruption in resource-rich countries? *World Development*, 37(3), 521–532.
45. Kolstad & Wiig (2009).
46. Kolstad & Wiig (2009).
47. Williams, A. (2011). Shining a light on the resource curse: An empirical analysis of the relationship between natural resources, transparency, and economic growth. *World Development*, 39(4), 490–505.
48. Abaza, H., Bisset, R., & Sadler, B. (2004). *Environmental Impact Assessment and Strategic Environmental Assessment: Towards an Integrated Approach*. Nairobi, Kenya: United Nations Environment Program.
49. Inderberg, T.H., Eriksen, S., O'Brien, K. & Sygna, L. (2015) *Climate Change Adaptation and Development: Transforming Paradigms and Practices*. London: Routledge.
50. Haggett (2009).
51. International Council on Human Rights Policy (2008).
52. Caney, S. (2005). Cosmopolitan justice, responsibility, and global climate change. *Leiden Journal of International Law*, 18, 747–775; see also Neumayer, E. (2000). In defence of historical accountability for greenhouse gas emissions. *Ecological Economics*, 33, 185–192.
53. See Kruger, J. & Pizer, B. (2004). *Europe Goes to Market: Opportunities and Challenges with a Trail Blazing Trading System* (pp. 2–5). Washington, DC: Resources for the Future; and Mangis, J.K. (1999). The United States sulfur dioxide emissions allowance program: An overview with emphasis of monitoring requirements and procedures. European Environment Agency Technical Report No. 29, November. Copenhagen, Denmark: European Environment Agency.
54. Langniss, O. & Wiser, R. (2003). The renewables portfolio standard in Texas: An early assessment. *Energy Policy*, 31(6), 527–535.
55. See Taylor, M., Suckling, K.S., & Rachlinski, R.R. (2005). The effectiveness of the Endangered Species Act: A quantitative analysis. *BioScience*, 55(4), 360–367.
56. See Kerkvliet, J. & Langpap, C. (2007). Learning from endangered and threatened species recovery programs: A case for using the US Endangered Species Act recovery scores. *Ecological Economics*, 63(2–3), 499–510.
57. See Taylor et al. (2005).
58. See Noecker, R.J. (1998, January). *Endangered Species List Revisions: A Summary of Delisting and Downsizing. Preliminary Findings of Analysis of Endangered Species Recovery Plans*, January. Washington DC: Congressional Research Service.

59. See US Government Accountability Office (2006). *Preliminary Findings of Analysis of Endangered Species Recovery Plans* (GAO-06–463R 2006). Washington, DC: GAO.
60. Cedillo, C. (2011). *Better Access to Remedy in Company–Community Conflicts in the Field of CSR: A Model for Company-based Grievance Mechanisms*. The Hague, The Netherlands: Institute for Environmental Security.
61. See Ruggie, J. (2008). Human rights and transnational corporations and other business enterprises. Statement Before the 63rd Session of the General Assembly of the United Nations, October 27. New York, NY: UN; Coglianese, C. & Nash, J. (2008). Government clubs: Theory and evidence from voluntary environmental programs. Corporate Social Responsibility Initiative Working Paper No. 50. Cambridge, MA: John F. Kennedy School of Government, Harvard University; Ruggie, J. (2011, March). *Guiding Principles on Business and Human Rights: Implementing the United Nations "Protect, Respect and Remedy" Framework*, March. New York, NY: United Nations.
62. Ruggie (2011).
63. Amnesty International (2014, March). *Injustice Incorporated: Corporate Abuses and the Human Right to Remedy*, March. London, UK: Amnesty International.
64. Singer, P. (1993). *Practical Ethics* (2nd edn) Cambridge, UK: Cambridge University Press, p. 8.
65. Singe, (1993) p. 8.
66. Fridahl, M. & Linnér, B.-O. (2015). Perspectives on the Green Climate Fund: Possible compromises on capitalization and balanced allocation. *Climate and Development*. DOI: 10.1080/17565529.2015.1040368.
67. Brown, L.R. (2001). *Eco-economy: Building an Economy for the Earth*. New York, NY: Norton and Company, pp. 236–237.
68. Beck, U. (1992). From industrial society to the risk society: Questions of survival, social structure and ecological enlightenment. *Theory, Culture & Society*, 9, 97–123.
69. Smith, Z.A. (2009). *The Environmental Policy Paradox* (5th edn). Upper Saddle River, NJ: Prentice Hall.
70. Orr, D.W. (1994). *Earth in Mind: On Education, Environment, and the Human Prospect*. Washington, DC: Island Press, p. 172.
71. Baer, P., Fieldman, G., Athanasiou, T., & Kartha, S. (2008). Greenhouse development rights: Towards an equitable framework for global climate policy. *Cambridge Review of International Affairs*, 21(4), 649–669; Baer, P., Athanasiou, T., & Kartha, S. (2007). *The Right to Development in a Carbon Constrained World: The Greenhouse Development Rights Framework*. Heinrich Boll Foundation; Baer, P., Harthe, J., Haya, B., Herzog, A.V., et al. (2000). Equity and greenhouse gas responsibility. *Science*, 289(5488), 2287.
72. See International Council on Human Rights Policy (2008); Harris, P.G. (2011). Introduction: Cosmopolitanism and climate change policy. In P.G. Harris (ed.), *Ethics and Global Environmental Policy: Cosmopolitan Conceptions of Climate Change* (pp. 1–19). Cheltenham, UK: Edward Elgar; Howard, M.W. (2011). Sharing the burdens of climate change: Environmental justice and qualified cosmopolitanism. In P.G. Harris (ed.), *Ethics and Global Environmental Policy: Cosmopolitan Conceptions of Climate Change* (pp. 108–127). Cheltenham, UK: Edward Elgar.

73. Linnér, B.-O. & Rayner, S. (2015). Moving climate policy towards an innovation investment agreement. *Research Handbook on Climate Governance*. Cheltenham, UK: Edward Elgar Press.
74. Jones, M.T. & Fleming, P. (2003). *Unpacking Complexity Through Critical Stakeholder Analysis: The Case of Globalisation*. Sydney, Australia: Macquarie University.
75. Aligica, P.D. (2006). Institutional and stakeholder mapping: Frameworks for policy analysis and institutional change. *Public Organizational Review*, 6, 79–90; Simonsson, A.K.L., et al. (2012). Method development for identifying and analysing stakeholders in climate change adaptation processes. *Journal of Environmental Policy & Planning*, 14(3), 243–261.
76. De Leon, P. & Varda, D.M. (2009). Toward a theory of collaborative policy networks: Identifying structural tendencies. *Policy Studies Journal*, 37(1), 59–74; Dryzek, J.S. & Tucker, A. (2008). Deliberative innovation to different effect: Consensus conferences in Denmark, France and the United States. *Public Administration Review*, 68(5), 864–876; Hendriks, C.M. & Carson, L. (2008). Can the market help the forum? Negotiating the commercialization of deliberative democracy. *Policy Sciences*, 41, 293–313; Edelenbos, J. & Klijn, E.H. (2005). Managing stakeholder involvement in decision making. *Journal of Public Administration Research and Theory*, 16, 417–446.
77. Rockloff, S.F. & Lockie, S. (2004). Participatory tools for coastal zone management: Use of stakeholder analysis and social mapping in Australia. *Journal of Coastal Conservation*, 10, 81–92; Weible, C,M., Sabatier, P.A., & McQueen, K. (2009). Themes and variations: Taking stock of the advocacy coalition framework. *The Policy Studies Journal*, 37(1), 121–140.
78. Heidrich, O., Harvey, J., & Tollin, N. (2009). Stakeholder analysis for industrial waste management systems. *Waste Management*, 29, 965–973.
79. Dunn, W.N. (2008). *Public Policy Analysis: An Introduction* (4th edn). Upper Saddle River, NJ: Prentice Hall.
80. Anton, D.K. & Shelton, D.L. (2011). *Environmental Protection and Human Rights*. New York, NY: Cambridge University Press, p. 431.
81. The Center for People and Forests (2011). *Free, prior, and informed consent in REDD+: Principles and approaches for policy and project development*, February. Bangkok, Thailand: The Center for People and Forests.
82. Goodland, R. (2004). Free, prior and informed consent and the World Bank Group. *Sustainable Development Law & Policy*, 4(2), 66–74; UN Permanent Forum on Indigenous Issues (UNPFII) (2005). Report of the International Workshop on Methodologies Regarding Free Prior and Informed Consent and Indigenous Peoples. Document E/C.19/2005/3, May 16–17. New York, NY: UNPFII.
83. Goodland (2004).
84. Finer, M., Jenkins, C.N., Pimm, S.L., Keane, B., & Ross, C. (2008). Oil and gas projects in the Western Amazon: Threats to wilderness, biodiversity, and indigenous peoples. *PLoS One*, 3(8), 1–9.
85. Goodland (2004).
86. Cornwell, L. & Costanza, R. (1999). Environmental bonds: Implementing the precautionary principle in environmental policy. In C. Raffensperger & J.A. Tickner (eds), *Protecting Public Health and the Environment: Implementing the Precautionary Principle* (pp. 220–240). Washington, DC: Island Press.

87. US Congress (2014). *Financial responsibility* [law: 33 U.S.C. § 2716(d)]. Washington, DC: US Congress.
88. Cornwell & Costanza (1999).
89. Kritz, N.J. (1996). Coming to terms with atrocities: A review of accountability mechanisms for mass violations of human rights. *Law and Contemporary Problems*, 59(4), 127–152.
90. International Accountability Project. (2012). World Bank Inspection Panel. (Oakland, CA: IAP).NO LONGER ONLINE
91. Carrasco, E.R. & Guernsey, A.K. (2008). The World Bank's inspection panel: Promoting true accountability through arbitration. *Cornell International Law Journal*, 41, 577–626.
92. See also World Bank (1998). *The World Bank Inspection Panel: The First Four Years (1994–1998)*. Washington, DC: World Bank Group; and World Bank (2003). *Accountability at the World Bank: The Inspection Panel 10 Years On*. Washington, DC: World Bank Group.
93. Cronstedt, C., Chambers, R., Margolis, A., Rönnegard, D., Thompson, R.C., & Tyler, K. (2014). An international arbitration tribunal on business and human rights. London, UK: Business and Human Rights Resource Centre.
94. These six attributes are synthesized from Amnesty International, 2014; Skinner, G., McCorquodale, R., & De Schutter, O. (2013). The third pillar: Access to judicial remedies for human rights violations by transnational business. The International Corporate Accountability Roundtable, CORE, and The European Coalition for Corporate Justice; Taylor, M.B., Thompson, R.C., & Ramasastry, A. (2009). Overcoming obstacles to justice: Improving access to judicial remedies for business involvement in grave human rights abuses. September 11–12. Proceedings of the Conference Conflict, Commerce, and Access to Justice, Oslo, Norway.
95. Paavola, Adger, & Huq (2006).
96. Hahn, F. (1982). On some difficulties of the utilitarian economist. In A. Sen & B. Williams (eds), *Utilitarianism and Beyond* (pp. 187–198). Cambridge, UK: Cambridge University Press.
97. See Branch, A. (2007). Uganda's civil war and the politics of ICC intervention. *Ethics & International Affairs*, 21(2), 179–198; Glasius, M. (2009). What is global justice and who decides? Civil society and victim responses to the International Criminal Court's first investigations. *Human Rights Quarterly*, 31(2), 496–520.

7 Insights from Political Economy for Adaptation Policy and Practice

1. Orr, D.W. (1981). Problems, dilemmas, and the energy crisis. In K.M. Gentemann (ed.), *Social and Political Perspectives on Energy Policy* (pp. 1–17). New York, NY: Praeger.
2. Ruhl, J.B. (2012). The political economy of climate change winners. *Minnesota Law Review*, 97, 207–277.
3. Brown, K. (2011). Sustainable adaptation: An oxymoron? *Climate and Development*, 3(1), 21–31.
4. Berkes, F., Hughes, T.P., Steneck, R.S., Wilson, J., Bellwood, D.R., Crona, B., Folke, C., Gunderson, L.H., Leslie, H.M., Norberg, J., Nyström, M., Olsson, P.,

Österblom, H., Scheffer, M., Worm, B. (2006). Globalization, roving bandits and marine resources. *Science*, 311, 1557–1558.

5. Klein, N. (2008). *The Shock Doctrine: The Rise of Disaster Capitalism*. New York, NY: Picador.
6. Mulugetta, Y., Jackson, T., & van der Horst, D. (2010). Carbon reduction at community scale. *Energy Policy*, 38(12), 7541–7545.
7. Ostrom, E. (2009). Background paper to the 2010 World Development Report: A polycentric approach for coping with climate change. Report Prepared for the WDR2010 Core Team, Development and Economics Research Group, World Bank. Bloomington, IN: Indiana University.
8. See Victor, D.G. (2011). Why the UN can never stop climate change. *The Guardian*, April 4; and Victor, D.G. (2011). *Global Warming Gridlock*. Cambridge, UK: Cambridge University Press.
9. Barnett, J. & O'Neil, S. (2010). Maladaptation. *Global Environmental Change*, 20(2), 211–213.
10. Alaniz, R. (2012). Unsupervised recovery: Adaptation strategies by two NGOs in post-Mitch Honduras. In M. Hamza & C. Corendea (eds), *Climate Change and Fragile States: Rethinking* (pp. 112–130). Geneva, Switzerland: Studies of the University Research Counsel, Education Publication Series of UNU-EHS No. 16/2012.
11. Storbjörk, S. & Uggla, Y. (2015). The practice of settling and enacting strategic guidelines for climate adaptation in spatial planning. Lessons from ten Swedish municipalities. *Regional Environmental Change*, 15(6), 1133–1143.
12. Buanes, A., Jentoft, S., Karlsen, G.R., Maurstad, A. & Soreng, S. (2004). In whose interest? An exploratory analysis of stakeholders in Norwegian coastal zone planning. *Ocean and Coastal Management*, 47, 207–223.
13. Shearer, C. (2012). The political ecology of climate adaptation assistance: Alaska Natives, displacement, and relocation. *Journal of Political Ecology*, 19, 174–183.
14. Bunce, M., et al. (2010), Policy misfits, climate change and cross-scale vulnerability in coastal Africa: How development projects undermine resilience. *Environmental Science & Policy*, 13, 485–497.
15. Bunce et al. (2010).
16. Sovacool, B.K. (2012). Expert views of climate change adaptation in the Maldives. *Climatic Change*, 114(2), 295–300.
17. Sovacool, B.K. (2012). Perceptions of climate change risks and resilient island planning in the Maldives. *Mitigation and Adaptation of Strategies for Global Change*, 17(7), 731–752.
18. Sovacool, B.K. (2011). Conceptualizing hard and soft paths for climate change adaptation. *Climate Policy*, 11(4), 1177–1183.
19. Adger, W.N., Paavola, J., & Huq, S. (2006). Toward justice in adaptation to climate change. In W.N. Adger, J. Paavola, S. Huq, & M.J. Mace (eds), *Fairness in Adaptation to Climate Change* (pp. 1–19). Cambridge, MA: MIT Press.
20. Raleigh, C. (2010). Political marginalization, climate change, and conflict in African Sahel States. *International Studies Review*, 12, 69–86.
21. Deressa, T.T., et al. (2009). Determinants of farmers' choice of adaptation methods to climate change in the Nile Basin of Ethiopia. *Global Environmental Change*, 19, 248–255.

22. Bryceson, D. (2002). The scramble in Africa: reorienting rural livelihoods. *World Development*, 30(5), 725–739.
23. Carr, E.R. (2008). Between structure and agency: Livelihoods and adaptation in Ghana's central region. *Global Environmental Change*, 18, 689–699.
24. Barrett, S. (2013). Local level climate justice? Adaptation finance and vulnerability reduction. *Global Environmental Change*, 23(6), 1819–1829.
25. Taylor, M. (2014). The political ecology of climate change adaptation: Livelihoods, agrarian change, and the conflicts of development. London, UK: Routledge.
26. Taylor (2014).
27. Verhoeven, H. (2011). Climate change, conflict and development in Sudan: Global Neo-Malthusian narratives and local power struggles. *Development and Change*, 42(3), 679–707.
28. Paavola, J. (2004). Livelihoods, vulnerability and adaptation to climate change in the Morogoro region, Tanzania. Working Paper EDM 04–12. Norwich, UK: Centre for Social and Economic Research on the Global Environment, University of East Anglia.
29. Woolgar, S. (1988). Reflexivity is the ethnographer of the text. In S. Woolgar (ed.), *Knowledge and Reflexivity: New Frontiers in the Sociology of Scientific Knowledge* (pp. 1–13). London, UK: SAGE.
30. Klein, R.J.T. & Juhola, S. (2014). A framework for Nordic actor-oriented climate adaptation research. *Environmental Science & Policy*, 40, 101–115.
31. Scheraga, J.D. & Grambsch, A.E. (1998). Risks, opportunities, and adaptation to climate change. *Climate Research*, 11, 85–95.

Index

www.ingramcontent.com/pod-product-compliance
Ingram Content Group UK Ltd.
Pitfield, Milton Keynes, MK11 3LW, UK
UKHW020837290926
15470UKWH00004B/1399

9781137496720